S0-CKQ-792

JUDGMENTS UNDER STRESS

Judgments Under Stress

KENNETH R. HAMMOND

New York Oxford
Oxford University Press
2000

Oxford University Press

Oxford New York
Athens Auckland Bangkok Bogotá Buenos Aires Calcutta
Cape Town Chennai Dar es Salaam Delhi Florence Hong Kong Istanbul
Karachi Kuala Lumpur Madrid Melbourne Mexico City Mumbai
Nairobi Paris São Paulo Singapore Taipei Tokyo Toronto Warsaw

and associated companies in
Berlin Ibadan

Published by Oxford University Press, Inc.,
198 Madison Avenue, New York, New York 10016

Library of Congress Cataloging-in-Publication Data
Hammond, Kenneth R.
Judgments under stress / Kenneth R. Hammond.
p. cm.
Includes bibliographical references.
ISBN 0-19-513143-6
1. Decision making. 2. Judgment. 3. Stress (Psychology)
I. Title.
BF441.H27 1999
153.8'3—dc21 99-10855

9 8 7 6 5 4 3 2 1

Printed in the United States of America
on acid-free paper

For My Family

Preface

This book puts forward a new approach to understanding the effects of stressful conditions on human judgment. My belief in the need for a new approach was precipitated after I undertook an extensive review (otherwise not available) of the literature created over the past 20 years. My conclusion is that the topic has been studied with a variety of theories, methods, and hypotheses that have never been organized into a coherent body of knowledge. I do not believe that anyone who tries to make use of these results for the purpose of training, the design of equipment, or human-environment interaction in general can do so now or will ever be able to do so; the work is internally incommensurable. I do not mean to say that many excellent studies have not been carried out; indeed, many have been, and there will be more. What is lacking is a general organizing principle that permits the development of useful knowledge. The purpose of this book is to provide such a principle.

Part I provides the reader with background with respect to the topic of judgment under stress. I begin by introducing the reader to this topic, considered so important that it was discussed in a Congressional hearing in 1988. I then broaden the discussion by describing the link between emotion and reason as it is understood today. That topic is followed by a brief description of the literature on the effects of stress on judgment. I found this literature unsatisfactory because it lacks any organizing principle and dismiss it as unhelpful. (Details of studies are provided in an annotated bibliography included in an Appendix.) I take two steps to remedy this situation. First, I provide an overall conceptual framework that describes the two metatheories that control research in the field of judgment and decision making, and the leading theories and research programs within each metatheory. I indicate the significance of each for the study of the effects of stress on judgment. Thus, part I provides the reader with the background that makes it possible to understand the need for a new approach.

The purpose of part II is to provide that new approach. I do so by providing a theory of stress and its origins (not merely a new definition), a theory of judgment that is explicated within the two metatheories described in part I, a theory of task-cognition interaction (because no such theory exists now), and a theory of task-cognition interaction under stress. In chapter 9 I take the risky step of extending the analysis from professional judgments to moral and political judgments under stress. I do so because I believe that the conceptual framework offers a reasonable step toward the improvement of our understanding of these judgments as well as professional ones. It is my view that the classic texts offer a view of morality that is too abstract, too general, or in any case too full of contradictions, to offer standards, or a helpful guide for the professional, moral, and political judgments of everyday life. My interest will be to work out a conceptual framework wholly within a psychological, not philosophical, context. Conclusions are presented in chapter 10.

Usage of the term "stress" does not become apparent in behavioral research before World War II, but the idea was certainly recognized by adventurers long before it appeared in psychology. The first use that I find is by the famous explorer Sir Ernest Shackleton, who led a party of 27 men over the Antarctic ice pack (1914–1916) in a failed attempt to cross the Antarctic continent. In her fascinating account of this journey, Caroline Alexander quotes from Shackleton's diary in which, at a particularly difficult decision point, he writes: "Am anxious. . . . Everyone working well except the carpenter: I shall never forget him in this time of strain and stress" (p. 113).

I should warn the reader about two features of my approach. One is that I do not immediately offer a definition of "stress." But without a definition, how can the reader know what I am talking about? Readers need only be patient; definition of stress is a matter that has bedeviled this topic from the beginning. I address this matter directly after presenting background material. Similarly with the topic of judgment; the reader will want to know exactly what I mean when I use that term. I meet that problem after reviews of the field in earlier chapters. I provide this background material so that the reader will see why I take the approach I do in part II. My goal is to present definitions grounded in theory, rather than arbitrary assertions.

The second feature that may surprise the reader is that I rely heavily on "case studies" or even anecdotal material (as well as standard experimental psychology). That is a serious matter. Let me explain why I took this path.

By any standard, the definition of stress must refer to judgments made under very difficult, even life-threatening, conditions, if it is to address seriously our subject. Simply subjecting subjects to loud noise, or other discomforts, will not convince many that the results obtained will tell us much about judgments made by persons in seriously stressful situations. And, as everyone knows, seriously stressful conditions are very difficult, if not impossible, to simulate in the psychological laboratory. And even if such simulations are carried out (very few have been), their fidelity is always suspect—the subjects know that nothing really bad is going to happen to them; in short, the essence of the topic is missing. In any event, such experiments as have been done are described in the bibliography in the Appendix, and readers may decide for themselves.

But this situation is not new. In the first book on "psychological stress," published 40 years ago, Janis (1958) began his *Preface* in this way:

> Various attempts to produce a "propositional inventory" of warranted scientific generalizations about stress behavior yield only a very meager list. About ten years ago, when I first began surveying the literature in this field, I became acutely aware of the lack of cogent, dependable evidence. There were, of course, many controlled laboratory experiments purporting to deal with stress behavior, but almost all of them dealt with extremely brief exposures to threat stimuli or measured only peripheral aspects of emotional excitement. The main source of difficulty, in my opinion, was that these carefully executed experiments had been carried out prematurely, before the significant variables in human stress behavior were adequately identified. Such experiments provide behavioral data which are generally quite reliable but of dubious value for extrapolating to the conditions of actual life stress.
>
> In contrast to the tangential laboratory investigations were a large number of field studies of major disasters, focusing on the effects of prolonged exposure to powerful stress stimuli. But most of these studies proved to be extremely weak in precisely those respects where the laboratory studies were strong: They consisted mainly of impressionistic accounts interpreted by observers who had failed to make use of any systematic procedures for minimizing the biasing influence of their own *a priori* expectations, attitudes, or emotional blind spots. (p. viii)

It is remarkable—and sad—that we can see that 40 years ago Janis put his finger on exactly what currently stands in the way of progress in scientific psychology when he said "most of these [field] studies proved to be extremely weak in precisely those respects where the laboratory studies were strong." That same difficulty confronts us today. On the one hand, we have rich, complex accounts of behavior that we can't use because of their unreliable nature, and on the other hand, we have highly reliable accounts of behavior occurring in the laboratory experiments that we can't use because its artificial character makes it of "dubious value for extrapolating to the conditions of actual life stress." Those who actually make judgments under stress will place their faith in results obtained from field studies or anecdotes—they can compare that information with their own experience—whereas researchers who have chosen to invest their lives in laboratory research will demand the assurance that scientific methods afford. And this divergence in choice—with its irreconcilable clash in values—will persist until someone, somehow, finds a way to encompass the advantages of each and diminish the disadvantages of each. I will try to do my part in later chapters.

There is one difference between the current situation and the one that Janis described, and that is there are now many reliable documents that afford good descriptions of events, in which individuals have been forced to make judgments under very difficult, life-threatening circumstances. (John Flanagan was a pioneer in this effort; see "The Critical Incident Technique," 1954.) Such documents offer high-fidelity material that can never be convincingly reproduced in a laboratory setting. The material in them is based on behavior of great interest, and therefore is extremely valuable; it should not be thrown away. Indeed, we need more good documents. We should learn as much as possible from them, all the while acknowledging their weaknesses, and being careful not to over-interpret what we find. The greatest danger in using such material is that it might be used as proof of an author's hypothesis. As we all know, however, single instances of reported behavior are no more than that; they prove no more than that such an event did happen, and often deceive us even then. That is a danger I have tried to avoid; I do not offer such material to prove anything other than the potential usefulness of a concept or theory to explain an event after the fact. That is, I have used documents to *illustrate* the meaning and relevance of a theoretical statement or concept. Naturally, I hope that my illustrations do just that—indicate that my theoretical statement or concept does have meaning and relevance for the topic of judgment under stress, and does contribute to, perhaps even increase, our understanding of that topic. Of course, I hope that, once the meaning and relevance of these ideas have been established in a context of interest, imaginative and enterprising investigators will develop experimental situations that represent judgment and decision making under stress, if that becomes possible.

There is another side to this choice to use documents in the attempt to explicate meaning, and that is that I find myself increasingly disillusioned with the application of the experimental method to the study of behavior. I recognize the rather beautiful logic of the design of experiments, and I fully appreciate the hard work that goes into carrying out experiments in order to be as responsible as possible to one's peers and to the public that supports the work. I have done my share of it. My disenchantment grows out of my increasing doubts about the use of what comes down to the experimental-control group method to provide us with anything but information about specific events in specific circumstances; the same charge that is laid against the anecdotal method. I have long argued that psychology has never come to grips with the fact that it is supposed to provide generalizations over conditions as well as subjects, and, as a result, has failed to provide us with general and significant findings (e.g., Hammond, 1948, 1954, 1966, 1996).

This is not a new criticism of experimentation in psychology. It was made earlier with some vigor during a period when academic psychology bravely took on a painfully honest self-examination of its recent past and found it wanting. This was done after one of the most determined efforts at lawseeking ever made by experimental psychologists, by a group of skilled, energetic, talented academics convinced that they had found the way to discover those laws. The review was carried out under the leadership of Sigmund Koch, 1959–1963, in what can only be described as an admirable fashion, yet steadfastly ignored, at least until 1998, when the six volumes that were produced were reviewed in a "Retrospective Review" in *Contemporary Psychology* (see Wertheimer, 1998). As Wertheimer noted, the questions raised as a result of that endeavor were described as "devastating." Furthermore,

> The profound overall conclusion of the entire study, which was doubtless radically contradictory to the expectations of the APA's Policy and Planning Board's Steering committee that had launched the project and of the individuals responsible for the National Science Foundation's support of it, was that psychology must "work its way free from a dependence on simplistic theories of correct scientific conduct." (p 9)

That harsh conclusion was also ignored; no committee was set up to inquire into the implications of that hard won statement.

Similar devastating criticisms, together with demonstrations of the consequences of such "simplistic theories of correct scientific conduct," had been made by the psychologist Egon Brunswik. But he was virtually alone; challenging the conventional wisdom in its own day was difficult and dangerous, and his words were for the most part also ignored and occasionally denounced. (This book does not ignore them.)

Psychology is not the only field that has suffered from the overzealous application of the experimental-control group paradigm. Although it has long been obvious that clinical psychology has been able to derive only meager benefits from the results of experimental-control group research, it has generally been supposed that this has been due to the clinicians' ineptness or unwillingness to seek the benefits from experimental psychology. In fact, the failure of transfer occurs, not because clinical psychologists are inept, but because this research paradigm is incapable of producing results easily transferred to practice. For it is not only clinical psychology that suffers from this inability to transfer results from lab to patient, clinical medicine suffers as well. (An example of how medicine is trying to come to grips with the problem of transferring research findings from clinical trials to the bedside of the patient can be seen in the symposium titled "Improving Evidence-Based Decision Making: Recent Advances in Research-Transfer Models" [Llewellyn-Thomas & O'Connor, 1998]; see also my discussion of D. Sackett's efforts to cope with this problem in Hammond, 1996.) Finally, and partly as a consequence of the gulf between "basic" and "clinical" psychology, I make no claim that the material in this book will be of use to clinical psychologists who must cope with the trials and tribulations of persons who seek help because they are "stressed out." I leave that situation to those better prepared to cope with it.

In sum, this book presents an unorthodox, and limited, approach to its topic; this preface is intended to prepare the reader for that.

I am indebted to Michael Drillings of the Army Research Institute for his personal help, and to the Army Research Institute (Contract No. DASW01-96-M-3143) for its support in the preparation of this book. The views, opinions, and/or findings contained in this book, however, are those of the author and should not be construed as an official Department of the Army position, policy, or decision. I am also indebted to Captain Paul X. Rinn (U.S. Navy, Ret.) for his comments on his remarks in the Congressional hearing and on an early draft of the manuscript. I thank Leonard Adelman, Michael Doherty, Bo Earle, Philip Laughlin, Leon Rappoport, Robert Schneider, and Thomas Stewart for their invaluable help in reading and criticizing early drafts of the manuscript. And I acknowledge with gratitude the excellent assistance of James Doyle, who actively participated in writing some sections, as well as that of Vivian Schneider and Mary Luhring in the preparation of this book.

Contents

I Where We Stand

II New Directions

Part I

WHERE WE STAND

1

What Do We Know About Judgment Under Stress?

Because I want to provide the reader with a broad overview of the study of stress on judgment and decision making as it has recently been carried out by psychologists and others, I will provide readers with some context that will give them an appreciation of where we stand today. Therefore, I begin with a description of a hearing held by the Defense Policy Panel of the Committee on Armed Services of the U.S. House of Representatives on August 3, 1988. It will be a crucible for our current understanding of the effects of stress on judgment and decision making.

A Congressional Hearing

A hearing was called by the House Committee on Armed Services to consider "the Administration proposal to pay compensation to the victims of Iran Air Flight 655 which was shot down [by the American cruiser *Vincennes*] over the Persian Gulf on July 3" (Committee on Armed Services, 1989, p. 1). Subsequent to the initial inquiry regarding compensation, on October 6, 1988, Les Aspin, chairman of the House Committee on Armed Services (later to be secretary of defense for a brief period in the Clinton administration), called a second hearing "to examine the impact of human factors such as stress" on the crew's performance, which, the chairman said "raised some interesting issues" (p. 189). Among the "interesting issues" identified by the chairman were two that are intrinsic to the principal question pursued here, namely: (1) "Does the performance during the shootdown [of Iran Air 655] identify aspects of human behavior that are poorly understood?" and (2) "What have researchers uncovered to date on man's ability to make rapid and even complex decisions in high-stress environments?" The chairman then observed that "to help explore these questions, we have a very distinguished panel of behavioral scientists gathered with the help of the American Psychological Association." In addition, the chairman noted: "We have Dr. Steven Zornetzer from the Office of Naval Research and Commander Paul X. Rinn, former commander of the USS *Samuel B. Roberts*" (p. 189). As a result, this hearing, and the episode that evoked it, have been the subject of many articles and discussions (see, e.g., Driskell & Salas, 1996), and will, no doubt, be the subject of many more. And although no one at the hearing said so, it is clear that the focus of the hearing would be on *professional judgments.*

The four behavioral scientists who testified were indeed "very distinguished," primarily for their work in the field of judgment and decision making, and therefore their statements to the panel provide an opportunity to discover what "researchers [have] uncovered . . . on man's ability to make rapid and even complex decisions in high-stress environments" (p. 189) in particular. In addition, Steven Zornetzer, director of the Life Sciences Directorate at the Office of Naval Research (ONR), had direct access to the research reports made over the past decade by (some of) these scientists as well as many others to the ONR regarding research on judgment and decision making. Thus, Zornetzer's testimony to the committee would be informed as a result of what ONR had learned during the (roughly) 10 years it had supported such research. Moreover, Commander Rinn would offer his conclusions regarding this topic. His conclusions would not be based on research, but on his direct experience with stress and judgment and decision making during recent combat in the Persian Gulf, in which his ship suffered explosion and fire.

Psychologists' Testimony

First, it is noteworthy that two of the four experts (Richard Nisbett and Richard Pew) did not cite research that relates stress and judgment and decision making. Nisbett focused on the fallibility of cognition in general, and Pew focused on questions related to decision aids and other aspects of decision systems. If it seems odd that two of the four expert witnesses did not address the topic of the hearings, the reason emerges from the testimony of the other two witnesses.

Of the two experts who did address the question put to them by the chairman, Robert Helmreich pointed out that "the whole area of stress is one that has been understudied" and observed that "we know little about stress in group situations" (Committee on Armed Services, 1989, p. 230). And Paul Slovic pointed directly to the absence of scientific knowledge about the topic, thus: "It is rather surprising to see how few studies have examined the effects of stress. . . . There are only a handful of laboratory studies that manipulate stress and observe the effects on complex judgment and decision making tasks. Most of these have employed time pressure as the source of stress" (p. 196). In his written testimony, Slovic (p. 200) reiterated the need for research on stress and judgment and decision making. Again, later in his written testimony he stated that it was "astounding to see how few studies have examined the effects of stress" (p. 209) and noted that a 1988 Numerical Aerospace Simulation (NAS) Committee recommended to Congress that such research should be undertaken (p. 209). Thus, it is not surprising that two of the four expert witnesses had nothing to say that bore directly on the issue at hand.

Slovic gave credit to work by Janis and Mann (1977): "Perhaps the most detailed theoretical treatment of stress in decision making has been provided by Irving Janis and colleagues (e.g., Janis & Mann, 1977)" (Committee on Armed Services, 1989, p. 209). He made no reference to the substance of the work by Janis and Mann, however. He also gave passing recognition to research on "human factors" when he pointed out the importance of the relation between "display features and . . . response structure."

Consistent with his frank statements about the "astounding" absence of research on the effects of stress on judgment and decision making, Slovic listed only six studies that have been done. He did claim, however, that "several consistent findings have emerged from these studies": "Under time pressure, the decision maker adopts a simpler mode of information processing. Rather than evaluate alternative actions completely, weighing and making tradeoffs among all the relevant attributes of each option, attention is focused on the one or two most salient cues and these tend to determine the decision" (p. 210). Further, "negative information gains in importance under time pressure. If the situation involves risk, time pressure leads to more cautious, risk-avoiding behavior, with greater importance given to avoiding losses" (p. 210).

Slovic concluded his review of research by stating: "In sum, studies of decision making under stress have uncovered important and consistent patterns of degraded information processing." But his conclusion was followed by the qualification that "research is needed to determine whether other forms of stress have effects similar to those of time stress to determine whether different types of judgment and decision making are more or less susceptible to the effects of stress, and to determine ways to reduce those deleterious effects" (p. 211). As we shall see below, his qualification is important.

Slovic's conclusions regarding the effects of (time) stress on judgment and decision making provided the only documented empirical research that was offered by the four experts. Nevertheless, taken as a whole, the testimony by the four experts in judgment and decision making gave a definite answer to the first question asked by the committee chairman ("Does the performance during the shootdown [of Iran 655] identify aspects of human behavior that are poorly understood?"); the answer given by the two experts who addressed the question was obviously "yes; judgment and decision making under stress" is an aspect of human behavior that is poorly understood. The answer to the second question ("What have researchers uncovered to date on man's ability to make rapid and even complex decisions in high-stress environments?") was not made clear, other than perhaps to suggest that no generalization would be supportable.

Nevertheless, generalizations about complex decision making "in high-stress environments" *are* made by those whose administrative positions (and scientific background) require them to respond to the demand for such generalizations. For example, based on what he had learned from a decade of ONR-supported research, Zornetzer offered a positive, unequivocal, generalized answer to the second question: "One of the things that happens under stress, for example, is the focus of attention shrinks. You tend to ignore more and more. You tend to try to focus in on just the critical issues, just the critical elements you need to get by to the next moment" (p. 391). Although Slovic's remarks (p. 210) emphasized much the same conclusion, albeit in more technical terms, this generalization is not supportable, as will be shown below. (Unfortunately, the experts were not cognizant of a fairly substantial literature on stress and decision making; for example, none seemed to be aware of Broadbent's 1971 well-known *Decision and Stress* which introduced the "filtering" hypothesis.)

Consensus Among the Experts

There were three conclusions on which the four experts agreed. (There were no disagreements among them.)

1. Paucity of research. Two of the four experts from the field of judgment and decision making emphasize the absence of research on the effects of stress on judgment and decision making, and the other two implicitly do so. But the Appendix of this report shows that there are many studies of judgment under stress reaching back to the 1970s.

2. The competence of human judgment is decreased by stress. All four emphasized the fallibility of human judgment, and despite the absence of substantial empirical evidence, all four indicate, more or less directly, that stress will have a deleterious effect on this already untrustworthy process. (They were wrong in this conclusion also; it is still uncertain whether stress generally has a deleterious effect.)

3. "Stress narrows the focus of attention." Only Slovic and Zornetzer were specific about the effects of stress; both indicated that the "focus of attention shrinks." Slovic was explicit about the negative consequences of the narrowing of focus, whereas Zornetzer implied that the consequences would be negative. (They were wrong; under the conventional

definitions of stress, narrowing may or may not occur, and such narrowing may or may not produce negative consequences.)

Each of my comments is explained below.

PAUCITY OF RESEARCH?

Zornetzer's documentation of the research supported by ONR did not list a single study of the effects of stress on judgment and decision making (Committee on Armed Services, 1989, pp. 260–271). One must then ask: How confident could he be that "attention shrinks" under stress? The answer must be "not very"—unless he had other sources in mind that he did not cite. In fact, all the experts, Zornetzer included, could have cited at least 25 studies that deal with the following stressors: sleep loss, shock, dangerous environments, time pressure, unrepresentative training, fatigue, memory/workload, noise, accident threat, heat. These and other studies are listed in the annotated bibliography in the Appendix.

The reason Zornetzer and the expert witnesses were unaware of the many studies of the effects of stress on decision making is that stress research is scattered through the clinical, social, human factors, physiological, and medical literature and is very uneven in scope and quality.

COMPETENCE OF HUMAN JUDGMENT DECREASED BY STRESS?

As long ago as 1976, Poulton had challenged the conventional view that stress (induced by stressors other than time pressure) always degraded performance (1976a). The title of his article made his point very well: "Arousing Environmental Stresses Can Improve Performance, Whatever People Say." Poulton stated, "There are well known rules that heat, noise, and vibration degrade performance. Yet a number of experiments show that all three stresses can reliably *improve* [italics added] performance, especially in tasks requiring speed or vigilance. . . . Experiments reporting improvements in performance need to be remembered as well as the experiments reporting degradations" (1976a, p. 1193). The possibility that stress might not degrade performance and may even enhance it was pointed out by Helmreich at the House Committee on Armed Services hearing (1989, p. 239). And there are many references to this outcome in the human factors literature as well.

When Slovic offered his generalization he was careful to note that the studies he cited included only time pressure (one among a long list of stressors he accurately cited). But even with respect to time pressure, there are contradictory results. For example, Rothstein (1986) found that "cognitive control (consistency of execution of a judgment policy) deteriorated under time pressure while cognitive matching remained unchanged" (p. 83). In other words, the attention span did not "shrink," the consistency of judgments did. And shortly after the hearings, Payne, Bettman, and Johnson had found that "people appear *highly adaptive* [italics added] in responding to changes in the structure of the available alternatives and to the presence of time pressure. In general, actual behavior corresponded to the general patterns of efficient processing" (p. 534). In view of these circumstances, we can hardly be confident in any generalization about the effects of time pressure. (Following the hearings, Svenson and Maule edited in 1993 a valuable book on *Time Pressure and Stress in Human Judgment and Decision Making* that I will discuss in later chapters.)

"STRESS NARROWS THE FOCUS OF ATTENTION"?

The generalization that one's "focus of attention shrinks" and that one "tends to ignore more and more" is not secure.

For example, Wickens (1987) states:

> The phenomenon of perceptual narrowing with arousal increase has received only few experimental demonstrations in more applied multicue situations such as the aircraft cockpit, or the industrial monitoring station, although anecdotal reports indicate that it is present there as well (Sheridan, 1981). One more applied context is the stress imposed by underwater diving. In a simulation of this hazardous environment, Weltman, Smith, and Egstrom (1971) found that a diver's ability to detect peripheral stimuli was impaired. It is important to note, however, that the phenomenon represents a mixture of optimal and non-optimal behavior. Arousal produces a non-optimal response by limiting the breadth of attention. But subject to this limit, *the human appears to respond optimally* [italics added] by focusing the restricted searchlight of attention on those environmental sources that are judged to be most important. (p. 64).

Wickens's remarks thus raise the question of whether it is, in fact, "deleterious" for the "focus of attention to shrink." Is it altogether a bad thing if under stress we "ignore more and more" and "focus in on just the critical issues"?

Commander Rinn's Testimony

Before proceeding further, it will be important and useful to turn to the testimony offered by Commander Rinn. His remarks were eloquent, and they were absolutely central to the topic of judgment and decision making under stress. Most important, they were in direct contradiction to the experts' contention that severe stress degrades the quality of judgment and decision making.

Differences Between the Experts and Commander Rinn

It is obvious from the printed record of the hearings that his testimony was of great interest to the committee. Several points stand out in Commander Rinn's testimony:

1. He was absolutely convinced that his *training* and the training given his crew members were excellent; it prepared him and them for the very stressful conditions they encountered when they found themselves battling a dangerous fire while their ship was taking on water after striking a mine in the Gulf. For example, "As a result of the training we received, . . . officers and crew of that ship were convinced that we could fight the ship successfully against all threats and we could save our ship if we had to" (Committe on Armed Services, p. 249).

2. He was convinced that training and preparation for combat (and crew selection procedures) were already of such high quality that further *research* on the effects of stress is not necessary.

3. He found no reason to believe that his judgment and decision making were *impaired* by the severe stress he experienced in attempting to keep his ship afloat. Commander (now Captain) Rinn described his decision to me in this way:

> I did not just focus on one problem and forget the others. I viewed the fire in the context that it could not sink me or destroy me. I had boundaries set to prevent its spread and although I could not put it out immediately it was not going to sink me within the next hour. The flooding however was going to send me to the bottom of the Persian Gulf very shortly if all efforts were not expended to stop it. The flooding would have put out the fires long before they exploded SAMUEL B. ROBERTS. (P. X. Rinn, personal communication, July 1993)

The details of the commander's story supported his claims about the value of training: "On the night of the 14th of April, when we received one of the largest explosions I have ever seen . . . , I can honestly tell you that the men of the ship reacted in some very impressive ways to the stress, to the danger, and to the catastrophic damage that we had" (p. 249). Evidence? They saved the ship. How was that done?

> We deviated from standard doctrine, but the important thing was that the members of my crew, when I told them to do that, and when I directed through the chain of command, carried out the orders and carried them out emphatically and executed the training that they had received, they didn't stammer, they didn't make mistakes, they didn't run the P250s the wrong way, but they did exactly what they had to do on the basis of the training they had gone through and because things had been made very clear to them how they had to function.
>
> To talk about other things happening in the face of the death there was an incident in a space called AMR2, which was one of my last main

> engineering spaces left that was not on fire and not flooding, I went into the space after about 1 hour and 30 minutes and my repair party was working in the space, one that we had worked very hard with to teach shoring electrical maintenance and also dewatering of spaces. If you can imagine stepping into a space that if you lose it, your ship is going to sink in 5 minutes and confronting enlisted men who range from an E-7 Petty Officer on down to basic seaman, and there are 12 of them in the space, and you are looking at a bulkhead that has four holes in it the size of a football—correction, a basketball—and the seam in the midst of that bulkhead is split, with water pouring in and you are standing in water up to your knees and the fire pumps you need to keep the ship afloat are only a foot above the water level and so are the main diesels that you are running to keep power to the ship that you need to get the ship out of the minefield, dewater the space and fight the fire, and you know that you are not really sure you are going to make it another hour, and you look at these enlisted men who you have trained through all of this training we have talked about and who you have stressed, but not necessarily stress like this, and you say, "This is not a very good situation, but the situation is this: We must save the space or the ship is lost. If you don't save that bulkhead and you don't save the space, you are not going to get out of here."
>
> They look at you, with great seriousness, with a small smile on their face and say, "Don't worry, Captain, we got this one in hand. In fact, this is nothing. You should see the next space."
>
> You go with them and they show you that space and in fact, there is a hole larger than the others and they are working frantically to close it. You quickly realize that these men have actually looked at the face of death and at the problem, and on the basis of the fact that they know how to do what they are supposed to do, they have confidence that they can succeed. It is only because of the training that they have and the confidence in the leadership. (pp. 250–251)

As a result, he and his men saved the ship.

In short, Commander Rinn strongly believes on the basis of hard experience that his training program made it possible for him and for his crew members to exercise good judgment, and that their cognitive abilities were unimpaired by severe stress. Even allowing for exaggeration born of pride in the performance of his crew, this view, based on direct experience, directly contradicts the implications of the testimony by the experts, namely, that stress makes what is already a poor quality process worse (see, especially, pp. 190–193 [Nisbett] and pp. 199–213 [Slovic]). Thus, Commander Rinn's testimony indicates that any laboratory-based research conclusion must withstand the criticism of those who have directly experienced the situations for which the laboratory results are intended. There is little doubt which conclusions will be accepted by consumers of the research.

WHY THE DIFFERENCE?

Conventional explanations are also traditional ones; the difference between military officers, business people, and all those whose expertise comes from experience and those scientists who explain behavior in terms of the results of their experiments has been with us for some time. (See Hammond, 1996 for examples.) Behavioral scientists have little patience for the explanations based on experience. Conclusions drawn from experience, they will say, are generally limited to poorly defined, uncontrolled conditions, tainted by wishful thinking, and biased by self-serving descriptions. None of these potentially fatal criticisms are difficult to find in after-the-fact reports.

But behavioral scientists' research is not difficult to criticize either; nonbehavioral scientists will point out that it produces fallible conclusions which have a short life, particularly when those conclusions are tested outside the laboratory. Such criticism is not new, and it even comes from behavioral scientists themselves. As long ago as 1943, the psychologists who created the research-oriented program for selecting spies for the newly formed Office of Strategic Services (OSS) at the beginning of World War II were determined to develop methods that would produce results applicable to the situations in which the spies found themselves. Here is the way those psychologists described their goals:

> Since most of the critical situations which were confronting the majority of OSS men in the field were both novel and stressful, we made our testing situations novel and stressful. Thus it may be said that the situational tests used at OSS assessment stations were as lifelike as circumstances permitted, incorporating some of the major components of situations that would naturally arise in the course of operations in the field. In other words, we tried to design assessment situations that would be somewhat similar to the situations in the management of which candidates would be judged by their superior officers and associates in the theater. (OSS Assessment Staff, 1948, p. 42)

Thus, in 1943 the psychologists on the OSS Assessment Staff saw the necessity for including in their research situations conditions *representative* of those in which their subjects would find themselves. Their research would have to justify itself in terms of its usefulness. The war had to be won. Psychologists continue to ignore what the OSS psychologists knew a half century ago, however; the need for the representative design of experiments still escapes them. As a result, the single testimony of an officer under fire can undermine the testimony of experts.

In short, both the behavioral scientists and the traditionalists have a point, and both are susceptible to criticism. The behavioral scientists still have to justify a methodology that struggles to produce results that generalize to the world outside the laboratory, and the nonbehavioral scientists still have to acknowledge the weakness of uncontrolled experience as a means of drawing conclusions about behavior. Fortunately, there are some signs that both sides are recognizing these problems.

Commander Rinn and Captain Haynes: Contrasting Situations, Contrasting Cognitive Activity

It will be useful to compare Commander Rinn's situation and his responses to it with those of another man who was in a similar situation—similar in that he was also responsible for the lives of several hundred people who, through no fault of his, in a matter of seconds were put in danger of sudden death.

On July 19, 1989, a DC-10, United 232 was on its way to Chicago from Denver when the flight crew suddenly realized that they had inexplicably lost the ability to control the aircraft; moving the controls had no effect on its direction or altitude; the aircraft was nearly out of control. The Voice Recorder shows Captain Al Haynes reporting that "we have almost no controllability," an astounding and incomprehensible event in the history of aviation. Radio appeals to UAL headquarters for advice were of no help; the manuals at the United base made no provision for an event of this sort. Yet something had to be done quickly or the airplane would crash and its hundreds of passengers might well lose their lives. The desperate nature of the situation can be seen in Captain Haynes's communication with the Sioux City (Iowa) airport control tower: "We have no hydraulic fluid, which means we have no elevator control, almost, none, and very little aileron control. I have serious doubts about making the airport. Have you got someplace near there, that we might be able to ditch? Unless we get control of this airplane, we're going to have to put it down wherever it happens to be." And a few moments later, he said, "We're trying to control it by power alone, we have no hydraulics at all, sir, we're doing our best, here." But what *could* be done? Surely, this was also a situation demanding judgment under stress, just as demanding as in the case of Commander Rinn on the *Samuel B. Roberts* in the Persian Gulf. The similarities and differences between the circumstances confronting Captain Haynes and those facing Commander Rinn form critical elements of our analysis, and therefore I explain them in detail.

At first glance, the similarities are clear; both men were forced to exercise their judgment knowing that death and destruction were imminent for hundreds of people for whom they were responsible unless appropriate action was taken quickly. As we shall see, however, there were sharp differences in both their *cognitive activity* and in the cognitive demands of their *situations*. Comparing them will let us become more specific, and introduce the reader to the manner in which both person and situation can be studied.

ANALYTICAL COGNITION IN RESPONSE TO STRESS

Consider first Commander Rinn on the bridge of the *Samuel B. Roberts.* His ship had struck a mine; there was an explosion, then fire, and then the ship rapidly began to sink. Here is how Commander Rinn described his situation in a letter to me:

> Fire is without a doubt the greatest threat aboard a ship at sea because if it sweeps through a ship there is no place to go and eventually you must stop it, abandon ship or die. On *Samuel B. Roberts* we were unable to extinguish the fires and that always increases the concern for explosion. However, as I assessed the situation aboard FFG 58 I knew we had physically limited its spread thus limiting the danger of explosion and progressive damage that would destroy us or drive us from the ship. The flooding on the other hand was rapidly sinking us at a rate of one foot every fifteen to twenty minutes and I knew I had little time left to stabilize that situation and stop progressive flooding before it sent *Samuel B. Roberts* to the bottom of the Persian Gulf. It was a balanced decision based on all the information known at that time, my training on how much damage my ship could take without sinking and my estimate of the situation based on years of operational experience. Although extremely stressed this decision was based on a logical judgment of which critical element would destroy my ship first. (P. X. Rinn, personal communication, July 1993)

I shall discuss Commander Rinn's judgments in more detail below; here I wish only to point out the *analytical* nature of those judgments. His training told him what he should think about, the rate at which the fire was destroying his ship, and the rate at which the water was sinking it. He thinks about these problems in quantitative terms, and he has the information that allows him to do that. It is clear he understands his situation very well, so well that he is prepared to act in opposition to naval doctrine; that is, he decides to dewater the ship first, and put out the fire second, although naval doctrine demands the reverse. Thus, his professional judgment ran head-on into a political judgment; if things go badly, his naval career will be over. One can see from his description of the situation, however, that his decision is plainly retraceable and defensible; he knows just what to do—and why he should do it—"although extremely stressed." His professional judgment—analytically derived—turned out to be correct; he saved the ship.

INTUITIVE COGNITION IN RESPONSE TO STRESS

Captain Haynes had a far different situation to cope with. Roughly five miles up in the air he encountered sudden and complete loss of control of the aircraft. Loss of control meant that moving the "yoke" (the wheel-like arrangement in front of him that moves the wings up and down and the nose of the airplane up and down) had no effect on the behavior of the aircraft. Pressure on the pedals at his feet that move the nose of the plane from right to left also were useless. In short, he could do nothing to control the aircraft; worse still, it began to take on movements of its own. And then luck intervened; a DC-10 pilot-instructor riding as a passenger on the plane realized that something was terribly wrong and sent a note via a flight attendant to the captain offering help. The captain accepted the offer and soon the instructor-pilot was the fourth man on the flight deck. He was given the task of manipulating the throttles that controlled the power to the engines. It was soon recognized that this task was critical; for the pilots discovered that by changing power to the engines on the wings, the direction and altitude of the aircraft could be roughly controlled. And that meant that it might be possible to land the plane and save its passengers. What it came down to, then, is that Captain Haynes and his crew—particularly the pilot at the throttles—had to find a new way to fly the airplane in order to bring it down safely. But they would have to do so without benefit of automation; kinesthetic, auditory, and visual cues would be the source of their information for controlling the orientation of the aircraft in the air and guiding it to a safe landing place. In short, the crew that, up to the point of the disruption, had been proceeding in a highly controlled, highly *analytical* cognitive mode, suddenly had to switch to perceptual-motor activity of a highly *intuitive* character. And they did so successfully; they landed the aircraft at an alternate airport, and although there was loss of life in the crash landing, most were saved.

ANALYSIS AND INTUITION: CONTRASTING RESPONSES TO CONTRASTING SITUATIONS

Thus we see that Commander Rinn solved his problem analytically, and Captain Haynes solved his intuitively. The point to be observed is that different stressful situations demand different cognitive activities; there is no universal means of coping with them. I shall come back to a discussion of the diametrically opposed—but successful—cognitive activity of these two men. First, however, we should look at the "stressors" they had to contend with. Were they the same?

Commander Rinn's description of the "stressors" operating on him is useful because we shall encounter these stressors later when we discuss research that has used various stressors in laboratory studies. In his case they included the threat of the loss of his life and/or severe injury to himself and the many members of his crew, and the distraction of the fire, heat, and noise in his surroundings. In addition, there was pressure for decisions to be made quickly, the necessity to control those immediately under his command, and the demand on his memory for what his training had taught him he must think about and do in such circumstances. The fact that Commander Rinn did not face a wholly *unexpected, unanticipated* event is significant, and that will be discussed below.

In Captain Haynes's case, he also suffered from the realization that in all likelihood he, his crew, and the passengers would shortly meet sudden death. Captain Haynes was spared the heat, the noise, and the general confusion we can assume that prevails when a ship is on fire and its ammunition is exploding. He was also spared, in a peculiar way, from another form of stress experienced by Commander Rinn, namely, lack of *training* for this event. No pilot had ever been trained for an emergency of this kind because no one had ever imagined such an emergency would occur. The loss of control of the aircraft of the type Captain Haynes experienced had never been anticipated by anyone. Thus, he did not have to search his memory for what his training had taught him to think about and do in an event of this kind for he had no such training that was specific to the loss of control of an aircraft. In a lecture Captain Haynes asks, "How did we do it? Not having any experience at all in flying an airplane under those conditions, our basic problem was keeping the airplane in the sky." So neither his many years of flight experience nor his training prepared him for the specific situation in which he found himself. Nor was the aircraft designed so as to provide him with resources for coping with this situation. He and his crew, including the flight attendants, and those on the ground did, of course, have training in coping with other emergencies, and that was of considerable benefit to UAL 232 at the time it hit the ground. But Captain Haynes never attended a training session on "Things To Do When You Lose Complete Control Of The Airplane," in contrast to, for example, the sessions he attended in relation to the loss of power in an engine. Of course, knowing that you don't know can be stressful as well; it leads to a feeling of helplessness that, in turn, can induce resignation and giving up.

This is in sharp contrast to Commander Rinn, who emphasized the critical importance of his training; that means his situation was anticipated, and the ship was designed for such an emergency. Nor did Captain Haynes have to worry about whether a large number of enlisted personnel would do what *they* had been trained to do. No one doubts, however, that Captain Haynes and his crew were "extremely stressed" (as Commander Rinn described his situation) and that the performance of the crew of UAL 232 was a remarkable achievement that will live forever in aviation history. And, of course, the uniqueness of the event made it unnecessary for Captain Haynes to worry about the political consequences of his actions. He did not have a naval doctrine to contend with, as did Commander Rinn.

Summary

The Congressional hearing was valuable in both its positive and negative aspects. It presented us with a clear view of the need for science-based knowledge (still lacking) about the effects of stress on human judgment.

It offered an unusual opportunity to see what experts in the field had to say about this topic. And it showed how little is known, and indicated how little is being done to improve our knowledge.

The hearing also offered an opportunity to see the difference in viewpoints between experts in research and theory and those who face the practical problems of making judgments under stress. As a result, I introduced the suggestion that different situations demand different forms of cognitive activity; some will call for increased analytical cognition, whereas some will call for increased reliance on intuition. Therefore, it is essential to describe the complexity of the situations people face as well as the complexity of their reactions to them, a conclusion that will be emphasized throughout this book.

2

The Obscure Link Between Emotion and Reason

New work by neuroscientists concentrates on the *interdependence* between emotion (associated with intuition) and reason (associated with analysis). That work holds interest for us for it is that interdependence that addresses the problem of the effects of stress on judgment. If emotion and reason are demonstrably interdependent in the brain, both functionally and structurally, then there will be no doubt that stress, insofar as it gives rise to emotion, will have some effect on judgment, inasmuch as judgment is a product of reason. First, however, we must consider the basic question of whether emotions would have a positive or negative effect on judgment, if the two were in fact interdependent.

The Role of Emotions in Judgment and Decision Making: Positive and Negative?

Lack of a coherent body of research has not prevented psychologists—and everyone else—from insisting that emotions play a significant role in the way decisions are made about important matters. The question remains, however, exactly what role do emotions play? And most important: Do they help or hurt? Do we want to make use of them, or get rid of them? This question will come as a surprise to most readers who will assume that we have always known that emotions are detrimental to good judgment, and that good judgment demands cool reason devoid of emotion. But recent pronouncements by established researchers offer a different view. For example, Richard and Bernice Lazarus, outstanding experts on emotion, picked up the following comment by the "respected and distinguished American diplomat George F. Kennan" (Lazarus & Lazarus, 1994, p. 198) that appeared in the *New York Times* of September 30, 1993 in which Kennan stated:

> There can be no question that the reason for . . . [the public's] acceptance [of the decision] lies primarily with the exposure of the Somalia situation by the American media, above all, television. . . . But this is an emotional reaction, not a thoughtful or deliberate one. . . . [It was] occasioned by the sight of the suffering of the starving people. . . . It is one which was not under any deliberate and thoughtful control. . . . I regard this move as a dreadful error of American policy. (Kennan, 1993, p. A25)

The psychologists took strong exception to this comment by Kennan; indeed, they told him he was wrong:

> The problem with this statement is that it pits emotion against reason even though the emotion charged with being the basis of the public acceptance—sympathy for a starving people—is itself the result of reason, albeit not the reason Kennan would like to have seen employed. It was, he believes, a poor judgment. But it is incorrect to see this reaction as any more emotional than if the decision had been made not to do anything . . . it is a careless—but common—usage to suggest that when we make bad decisions, they are based on emotion, but when we arrive at good decisions, they are based solely on reason. (Lazarus & Lazarus, 1994, p. 199)

That comment could draw attention—perhaps derisive attention—from the diplomatic corps, if not from everyone. The diplomatic corps will be surprised to be charged with being "careless," for assuming that when they "arrive at good decisions, they are based solely on reason." Won't they ask one another: "Isn't that what we are supposed to do? Aren't we supposed to base our professional judgments solely on reason? How else can we reach good decisions? Are we expected to go before a Congressional committee and proudly announce that our reasoning, our judgment about what action should be taken, was guided by our *emotions*?" The psychologists' argument that emotion is *not* opposed to reason will appear incredible, possibly silly, in the eyes of many.

But is it? Doesn't it seem odd that Kennan wants us to appeal to reason alone, to explain, coolly, rationally, why it was *not* immoral for the people of one nation to let those in another starve, even though they had the means to prevent it? And why is it that sympathy is made to appear *thoughtless*, wrong, and dangerous, and therefore a reason why aid should have been denied. Was this judgment *entirely* a professional judgment, one in which emotions should be removed from reason altogether? Does the reader agree with that? Or does the reader believe that emotions *should* guide our reasoning? Or perhaps it is inevitable that, because of our biological makeup, emotions will *always* guide our reasoning, for better or worse. Psychologists argue about that, though few others do.

A somewhat different approach to the question of the relation between emotion and reason was taken by Mellers, Schwartz, Ho, and Ritov (1997). They conducted some very rigorous experiments within the framework of an equally rigorous theory. They conclude their article by noting that "regardless of what we ultimately decide to do about emotions and [reason], we still need a better understanding of emotional experiences, both actual and anticipated. Simply dismissing our emotions as irrational will surely leave us vulnerable to their effects." This assumes that emotions do have their effects, and in addition, it implies that we don't *want* to be "vulnerable to their effects."

Emotions Depend on Reason? Reason Depends on Emotion?

Does emotion depend on reason? Or does reason depend on emotion? Or are they independent of one another? These are at once profound questions and they lie at the source of this dispute. The Lazaruses say "yes" to the first question; tradition says "yes" to the second; but Kennan, and many others, demand that we act in accordance with the third. Therefore, we must look into these questions inasmuch as our topic is the effect of stress and thus emotion on judgment, and inasmuch as we are finding that the views of thoughtful policymakers are in conflict with those of equally thoughtful and responsible scientists.

Lazarus and Lazarus (1994) didn't stop with arguing that it is a mistake to pit emotion against reason. They went further to assert that "nowadays psychologists concerned with emotion are beginning to think that this view [that emotions should be separated from reason] is quite false—that, in fact, emotions *always depend* [italics added] substantially on reason" (p. 199). There are two points here: One is that emotions and reason are inseparable, the other is that the emotion we experience is dependent on our judgment of the situation; *first* we judge, *then* we experience emotion. Note that the argument has now shifted in an important way. The psychologists are no longer talking about what *ought* to be—whether reason *ought* to be separated from emotion so that our judgments can achieve rationality; they are now asserting what the *facts* are—or at least what they believe the facts are. That is, they argue that "emotions always depend substantially on reason" because they believe that research shows that is how these psychological processes actually work, and *why they work the way they do*. The answer to the "why" question brings us to research by neuroscientists who investigate the structure and function of the brain. They have an answer to these questions that is similar to that of the Lazaruses.

Neuroscientists: "Reason May Not Be So Pure"

The argument put forward by Lazarus and Lazarus is getting substantial support from those neuroscientists who carry out research on emotions and the brain, who seek to determine the structures of the brain that are associated with both reason and emotion. Antonio Damasio (1994) is one of these neuroscientists, and he says: "I propose that reason may not be as pure as most of us think it is or wish it were, that emotions and feelings may not be intruders in the bastion of reason at all: they may be enmeshed in its networks, for worse *and* for better" (p. xii). Note that Damasio acknowledges that emotions may work to our disadvantage ("for worse") and advantage ("for better"). But it is clear that he is suggesting something new and different, for he later says: "I suggest only that certain aspects of the process of emotion and feeling are indispensable for rationality" (p. xiii). Can this be true? Are emotions, or at least "certain aspects of the process of emotion and feeling *indispensable* [italics added] for rationality"? Can Damasio be serious? Tradition has it that emotion must be *dispensed* with if we are to achieve rationality, that is to say, defensible decisions.

Damasio's argument is not merely theoretical; he will support it with neuroscientific research, and he is by no means alone in his point of view. Although Joseph LeDoux (1996) is also a neuroscientist who, in his book titled *The Emotional Brain*, puts forward a somewhat different view than Damasio, he also wants to persuade us that thoughts and feelings are inevitably entangled. He does so for the same reason as the Lazaruses and Damasio; that is, the processes of evolution—natural selection—have designed the brain to work that way. The principal problem for the neuroscientists is to find the structures within the brain that account for the entanglement, and they are making progress in that effort. Now some of these researchers have gone beyond *acknowledging* that reason and emotions are inseparably entangled in everyday life to asserting that because they are functionally and structurally entangled, they should therefore *not be disentangled* in our theories about judgment and decision making, nor should they be disentangled in our research projects for fear of creating artificial conditions of no relevance to their topic.

But more startling developments have occurred. Damasio and LeDoux, for example, have gone so far as to argue that this entanglement not only *must* be so (because observations of the structure of the human brain show us that), but also that it is a *good thing* that it must be so. For at least on some occasions, our judgments and decisions will somehow be better when emotion and reason are *not* separated. This suggestion is startling because it is the reverse of our long-standing belief in the advantage of "cool reasoning" over impassioned reasoning. Nor are these researchers mere "armchair" theorists; they can point to experiments (and fascinating case histories) that offer evidence for their argument.

Important decision makers like George Kennan will have to listen to them.

But they may not, for it is hard to imagine two more disturbing suggestions; they fly in the face of the long history of one of our most widely accepted premises; they deny the wisdom of what everyone has been taught—namely, that reason be independent of our emotions—which assumes that they *can* be independent. In a word, you must "stay calm" when exercising your judgment. Exactly what George Kennan—and the reader—took for granted. Think of the Cuban missile crisis. Does anyone think that emotion ruled the day among those all-powerful (and all-wise?) committee members faced with that crisis? Does anyone think that emotion *must* have been part of the decision process? Does anyone think that the reasoning by the committee assembled by President Kennedy *should* have been dependent on their emotions? Some psychologists will say that the answer is "yes" to all three questions.

This bold—and unexpected—attack on the age-old premise that emotion should be—or can be—separated from reason comes from well-informed researchers. Suppose they are right. They have evidence for their argument, so they might be. Therefore, our topic—judgment under stress—requires that we examine what they have to say, and so I now turn to a brief discussion of each of these propositions.

Proposition 1: Emotions Are Inescapably Entangled With Reason

There are two types of arguments for this proposition: a functional one and a structural one. The functional argument is put forward by students of emotion, such as the Lazaruses. The structural argument—very recent—is put forward by neuroscientists such as Damasio (1994) and LeDoux (1996) and put in popular form by Goleman (1995). These three books, appearing at almost the same time, are certain to have a profound effect on the way scientists—and the public—think about judgment and decision making. A recent experiment by Damasio, for example, was described in the *New York Times* as "an experiment with broad implications for human behavior" (Blakeslee, 1996, p. C5).

The Functional Argument

Lazarus and Lazarus (1994) put the functional argument bluntly: "Our position in this book is that emotion depends on reason, and that *there is no way to separate them* [italics added]. To do so fails to recognize the role of reason in the arousal of emotions" (p. 203). These authors take the position that cognition necessarily comes before emotion; one "appraises" the situation, and having thus "reasoned," one's emotion follows. This position is diametrically opposed to that taken by another well-known student of emotions, Robert Zajonc, who, 14 years earlier, declared that just the opposite was true—emotion precedes cognition (Zajonc, 1980). Lazarus and Lazarus, however, make no mention of Zajonc's work, despite its frequent citation as a landmark article. These opposing positions have a rather long history, both of which are reviewed by LeDoux (1996).

But LeDoux, although a neuroscientist, comes down on the side of a special form of separatism: "Emotion and cognition are best thought of as separate but interacting mental functions mediated by separate but interacting brain systems" (p. 69). As may be seen from that remark, LeDoux supports his separatist functional argument with considerable reliance on a structural argument. But his separatist stand is weakened by his use of the term "interacting." Just what is "interacting" supposed to mean if it does not mean "interdependent"? Doesn't interacting imply that emotion *affects*, that is, changes, cognition? Yes, it does, or at least it implies that it *can*. But does it imply that emotion *depends* on reason? That is, does it imply that the emotion we experience will depend on our "appraisal" (Lazaruses's term) of a situation? It could imply that, and it could imply the opposite, that our "appraisal" depends on our emotions. So LeDoux's statement about the "interaction" between "brain systems" is less clear than the Lazaruses's statement. It is worth noting, however, that this question is far from being settled; there are many neuroscientists who take the separatist view and many who, like Damasio, do not.

The Structural Argument

The neuroscientists Damasio and LeDoux do support the proposition that emotions and cognition are entangled somehow, and that it must be so because the brain is built the way it is. LeDoux is very specific about this. He traces out pathways that involve the hippocampus and the amygdala and tells us that

> hippocampal circuits, with their massive neocortical connections, are well suited for establishing complex memories in which lots of events are bound together in space and time. The purpose of these circuits . . . is to provide representational flexibility. No particular response is associated with these kinds of memories—they can be used in many different ways in many different kinds of situations. In contrast, the amygdala is more suited as a triggering device for the execution of survival reactions. Stimulus situations are rigidly coupled to specific kinds of responses through the learning and memory functions of this brain region. It is wired so as to preempt the need for thinking about what to do. (p. 224)

Note the phrase "preempt the need for thinking about what to do."

Proposition 2: It Is Good *That Emotion and Reason Are Entangled*

In this proposition psychologists and the neuroscientists are persuaded that the entanglement works to our advantage. They believe that we are all better off because of this arrangement, which we owe to evolution. For example, the Lazaruses, relying on the work of the neuroscientists, state: "We are now confident that the frontal lobes of the cerebral cortex play a major role in the emotions, and that emotion and reason are interdependent—in effect, that there is extensive networking between the older, primitive brain and the newer, advanced brain" (1994, p. 179). And that is good because: "Emotions—and the capacity to think intelligently—evolved because *they facilitate survival* [italics added] and help us flourish" (p. 179). The Lazaruses think they do this by "mobiliz[ing] us, [thus] providing added strength and endurance in emergencies" and "when we experience an emotion, our mind concentrates its attention on the emergency and what might be done to cope with it" (p. 179). Further, emotions "give urgency to making distinctions quickly among safe, dangerous, and opportunistic situations" (p. 179). And finally, "Emotions have to be smart to facilitate survival, and thus intelligence and emotion evolved together in our species" (p. 180). It is clear that these psychologists believe that the putative interdependence of emotion and reason is a good thing because it enhances our chances for survival and thus has an evolutionary base.

Because Damasio is a neuroscientist, he is more focused than the Lazaruses on the specific neural mechanisms that are involved in the relation between emotion and reason. He notes that "some of the basic regulatory mechanisms [of our bodies] operate at covert level and are never directly knowable to the individual inside whom they operate. You do not know the state of the various circulating hormones, potassium ions, or the number of red blood cells in your body unless you assay it." But others you do know about because they "drive you to perform (or not) in a particular way. These are called instincts." (Damasio, 1994, p. 116). Damasio then describes how various changes in the regulation of physiological processes in the body affect the brain, which result in changes in behavior (e.g., changes in blood sugar affect the hypothalamus which leads you to feeling hungry and to eat; eating changes blood sugar level which, in turn, affects the hypothalamus and results in a cessation of eating). Damasio makes the point that these "preorganized mechanisms are important not just for basic biological regulation. They also help the organism classify things or events as 'good' or 'bad' because of their possible impact on survival." That is, "the organism has a basic set of preferences—or criteria, biases, or values" (p. 117).

So there is a clear interdependence between bodily events and the brain's classification of things and events in such emotive terms as good or bad. Modern science enables Damasio, and of course other neuroscientists, to trace out that interdependence in great detail. And because "good" or "bad" is interpreted as being good or bad for survival, we are thus led to an evolutionary explanation of why we are the way we are, much as the Lazaruses (above) indicated. Damasio is explicit about this connection:

> From an evolutionary perspective, the oldest decision-making device pertains to basic biological regulation; the next, to the personal and social realm; and the most recent, to a collection of abstract-symbolic operations under which we can find artistic and scientific reasoning, utilitarian-engineering reasoning and the development of language and mathematics. But although ages of evolution and dedicated neural systems may confer some independence to each of these reasoning/decision-making "modules," I *suspect* [italics added] they are all interdependent. (p. 191)

I have italicized the word "suspect" in that sentence from Damasio because it indicates that although he provides us with rather energetic arguments for the neurological and evolutionary basis for the interdependence of emotion and reason, apparently he is not sure that the question is settled. And as to whether this putative interdependence is a good thing, Damasio becomes equivocal once more. For example: "While biological drives and emotion may give rise to irrationality in some circumstances, they are indispensable in others . . . [especially] in the personal and social domains" (p. 192). And he makes his equivocation plain and justifies it by saying: "If you are wondering how bizarre it is that biological drives and emotion may be *both* beneficial and pernicious, let me say that this would not be the only instance in biology in which a . . . mechanism may be negative or positive according to the circumstances" (p. 194). In short, even though it may not be fully established, Damasio takes his interdependence seriously and finds that it may or may not be a good thing, depending on circumstances.

LeDoux traces out the interdependencies between cognitive and emotional processes, as well as their anatomical relations, as one might expect from a neuroscientist. As for the question of whether such interdependence is a good thing, we find him saying: "As things stand now, the amygdala has a greater influence on the cortex than the cortex has on the amygdala, allowing emotional arousal to dominate and control thinking" (1996, p. 303), thus contradicting the Lazaruses's claim that emotion depends on reason (see above). And "although thoughts can easily trigger emotions (by activating the amygdala), we are not very effective at willfully turning off emotions (by deactivating the amygdala). Telling yourself that you should not be anxious or depressed does not help very much" (p. 303). (One can't help wondering whether Commander Rinn and Captain Haynes did not successfully tell themselves that they should not be anxious or depressed.) But LeDoux adds a new possibility:

> The increased connectivity between the amygdala and cortex involves fibers going from the cortex to the amygdala as well as from the amygdala to the cortex. If these nerve pathways strike a balance, it is possible that the struggle between thought and emotion may ultimately be resolved not by the dominance of emotional centers by cortical cognitions, but by a *more harmonious integration of reason and passion* [italics added]. (p. 303)

That is indeed a compelling thought, and we shall return to a discussion of it in exactly those terms in relation to psychological theory and moral judgment at the Supreme Court (chapter 9).

At present then, we cannot depend on either theory or empirical fact to settle the question of the relation between emotion and reason, and that is an important conclusion for our approach to the study of the effects of stress on judgment. We can be sure that this question has absorbed the attention of scientists, both those who are students of emotion and reason, such as the Lazaruses, and those who are students of the brain and emotion and reason, such as Damasio and LeDoux and many others. We cannot, however, be certain of what the relation between emotion and reason is; for example, which is dependent on which, or whether there is any connection at all. And without that knowledge it is difficult to see what conclusions can be drawn about the effects of stress on judgment. At least part of the problem is that four different fields of investigation are isolated from one another.

Four Fields in Isolation

Irrespective of the fact that they are studying intuition and analysis, each of the four fields—emotions, neuroscience, evolutionary psychology, and cognitive science—has its own literature, its own set of concepts, its own professional society, and each enjoys carelessly generalizing its findings to the other fields. Neither students of emotion nor neuroscientists nor the biologists seem to be willing to include in their theorizing and research the literature of the field of judgment and decision making, with its thousands of empirical studies and several contesting theories, and its own professional society (which, regrettably, is in turn separate from the Cognitive Science Society). The fact of the existence of a literature of thousands of studies on judgment and decision making means that there is a long history of research that has tested a variety of methods and procedures used for the purpose of analyzing judgment processes—all ignored by the others. (Damasio does refer once, briefly, to the work of Kahneman and Tversky, but ignores everything else.) We cannot move ahead in our understanding of the effects of stress on cognition as long as separatism remains the standard paradigm.

The effects of separatism can be seen in the index of the book by Lazarus and Lazarus. They list "judgmental errors" and devote seven pages of text to "sources of erroneous judgments," but these pages do not include a single reference to research in the field of judgment and decision making; they suggest only that emotion causes the errors (which is a curious contradiction of their criticism of George Kennan). LeDoux also cites Kahneman and Tversky, but their work is given only a sentence or two. And although the index to Damasio's book contains the word "judgment," it is followed only by "see emotion." I think it would be fair to say that the neuroscientists' and emotion researchers' concept of a judgment task is about as sophisticated and undifferentiated as the judgment researchers' concept of the brain. And the biologists ignore this field altogether.

Acknowledgment of this situation is rare in the literature but commonplace in conversations. It can be seen, however, in the complaints of emotion researchers Monroe and Kelley (1995) regarding their colleagues' treatment of the concept of "appraisal." They state that "measuring an individual's appraisal of stress appears to be a very straightforward means of testing cognitive models of stress and their implications for health and well-being" (p. 122). But, they go on to note: "In this context it is rather bewildering that there are relatively few measures currently available." Monroe and Kelley's bewilderment is justifiable, for they find that although Lazarus (1991) has made appraisal a "central concept" in his work (e.g., "The central concept of my theoretical analysis of psychological stress is *appraisal*"), he has not differentiated appraisal from judgment. The reader should stop at this point and ask: What is an appraisal if not a *judgment*? Isn't that what Commander Rinn and Captain Haynes were doing—making an "appraisal," a judgment of a situation? If someone makes an appraisal of a situation, is that person not making a judgment of it? And if a psychologist does indeed want to make a distinction between an appraisal and a judgment, shouldn't he or she be explicit about that distinction? If Lazarus and his colleagues intend appraisal to mean what judgment means, then Lazarus could (and should) refer to the literature on judgment and decision making and link it to the study of emotions. And if something different is intended, they should say what the distinction is and why it is a necessary one.

Monroe and Kelley (1995) make this clear when they quote with approval from Grinker and Spiegel (1945), possibly the first users of the concept of stress in modern psychology: " '*appraisal* of the situation requires mental activity involving *judgment* [italics added], discrimination, and choice of activity based on past experience' " (p. 122).

But Monroe and Kelley are factually wrong, there are numerous cognitive models of "appraisal," that is to say, "judgment," available to emotion researchers and have been for decades, *if* you agree that the two words mean the same thing. The bewilderment of Monroe and Kelley occurs because Lazarus separated appraisal from judgment, and thus removed this "central concept" from the judgment and decision making literature and the possibility of making use of the cognitive models Monroe and Kelley are looking for.

Unfortunately, the reverse is also true; there are essentially no references to the neuroscience literature, or the literature on emotions or evolutionary biology by judgment and decision researchers. No one in the judgment and decision field that I know of has referred to the "appraisal" process. A similar problem exists between the judgment and decision researchers and the neuroscientists. The gulf between these disciplines becomes particularly obvious when the judgment tasks used by neuroscientists are examined; they are generally useless from the point of view of a judgment researcher because they fail to consider the task parameters essential from their point of view. And the rich material from the behavioral ecologists has been completely ignored altogether.

Nothing illustrates the failure of progress in understanding the effects of judgment and stress better than the failure to coordinate the work in the four fields that are responsible for advancing our knowledge of the topic.

Implications for the Study of Stress

Because of my conclusion that it is impossible to draw any general conclusions from the research that directly addresses the topic of the effects of stress on judgment, I do not now present a summary of a study-by study overview of that work. Rather I have provided in the Appendix an annotated bibliography that examines that body of work in detail. This material should offer the reader an opportunity to evaluate my conclusions more readily than a summary review that is bound to be selective. In addition, the bibliography is up-to-date at the time of this publication and probably the most extensive one available. I believe that a perusal of the bibliography will lend support to my view that the topic has been studied with a variety of theories, methods, and hypotheses that have never been organized into a coherent body of knowledge—nor can they be. No doubt many excellent studies have been carried out, but the lack of a general organizing principle defeats any attempt to make use of this body of work. And, indeed, there has been none. So I urge the reader to examine the bibliography and to test the validity of the conclusion I have drawn. For if there is, in fact, a principle that can be applied to draw this work together, it certainly should be applied. For my part, in the next chapter I put forward a conceptual framework that I believe will provide the organizing principle we so badly need.

3

Two Metatheories That Control Theory and Research in Judgment and Decision Making

Frameworks for describing a variety of theories and models are called *metatheories.* What is a metatheory? It is simply a means for describing and classifying different *types* of theories. Classifying various theories—and naming them—is useful because it enables us to see their differences in goals, and thus enables us to see why their content and their associated methodologies differ.

Coherence and Correspondence Metatheories

I will put the idea of metatheories to work immediately by pointing out that *two* metatheories have controlled the direction of research in judgment and decision making from its beginning in the 1950s. One is named the *coherence* metatheory because it refers to those theories and models that are concerned with the coherence—that is, the rationality of a person's judgments and decisions. Research carried out within these theories seeks to discover what factors affect whether a person's judgment process will meet the test of rationality, that is, logical or mathematical consistency, or in the case of a narrative, an absence of contradictions. A second metatheory controlling research in judgment and decision making is named the *correspondence* metatheory because it refers to those theories that are concerned with the correspondence of a judgment with the empirical facts, that is, the empirical accuracy of a judgment. Was the forecaster's prediction of rain correct? Did it in fact correspond with the occurrence of rain? Was the physician's prognosis correct? Did it correspond with the patient's recovery? Research carried out within the theories and models within the correspondence metatheory thus seeks to discover whether the prediction corresponded with the facts, and what factors lead to empirical accuracy or inaccuracy of judgment.

Both types of judgments are important, of course, but they need to be distinguished if we are to understand their nature. Therefore, classifying theories according to which type of judgment process is under study by a group of researchers will make it possible for us to understand why investigators working within one metatheory carry out different forms of research on judgment and decision making than those who work within the other metatheory. Those researchers who study the coherence of judgment processes and those who study the correspondence of judgment processes comprise two different groups because they have different interests; they focus on different cognitive processes.

For a striking illustration of this division of efforts, the reader should compare the books by R. Cooksey (1996) and E. Poulton (1994). Both purport to describe the field of judgment and decision making, both are responsible to their tasks, both are scholarly achievements, yet there is little overlap in content. No one new to this area could be blamed if they thought these were books about two different fields. They describe two different topics, name different researchers, cite different articles and books, address different issues, and bring forward different problems for analysis and discussion. Nor has it been apparent to theorists and researchers in the field that this is the case; hardly any mention of this disparity has ever been made in the literature (but see D. Frisch and R. Clemen, 1994).

Indeed, even the most sophisticated and experienced researchers have confused these two metatheories in the most fundamental way. For example, the reader should note how A. Tversky and D. Kahneman (1974) confused these two approaches when they introduced their work on cognitive "biases" as follows:

> The subjective assessment of probability resembles the subjective assessment of physical quantities such as distance or size. These judgments are all based on data of limited validity, which are processed according to heuristic rules. For example, the apparent distance of an object is determined in part by its clarity. The more sharply the object is seen, the closer it appears to be. This rule has some validity, because in any given scene the more distant objects are seen less sharply than nearer objects. However, the reliance on this rule leads to systematic errors in the estimation of distance. Specifically, distances are often overestimated when visibility is poor because the contours of objects are blurred. On the other hand, distances are often underestimated when visibility is good because the objects are seen sharply. Thus, the reliance on clarity as an indication of distance leads to common biases. Such biases are also found in the intuitive judgment of probability. This article describes three heuristics that are employed to assess probabilities and to predict values. Biases to which these heuristics lead are enumerated, and the applied

and theoretical implications of these observations are discussed. (p. 1124)

The authors do not explain to their readers that because they are discussing the subjective use of a visual cue ("clarity") with respect to its *empirical* validity, they are, therefore, discussing its use in determining the distance of an empirically real object, and, as a result, they are discussing *correspondence competence;* that is, the competence of persons to make correct distance estimates. That omission becomes critical, for at the end of their discussion of errors of estimation, they assert that "such biases are also found in the intuitive judgment of probability." But "such biases" are not found in the intuitive judgments of probability. The authors' error lies in failing to recognize that the "errors" in the two circumstances are very different. In intuitive judgments of the distance of an object, an error occurs when the person's subjective judgment of distance does not agree with an objective measure of distance. In the intuitive estimate of a probability, an error occurs when the estimate does not agree with the calculation of a probability from a mathematical formula. The probability estimation entails an analytical, justifiable formal process that serves as a criterion for the rationality of the subject's probability estimation. Indeed, getting the process correct is more important than getting the correct answer. The distance judgment, on the other hand, entails nothing of the sort; accuracy in the estimation of the empirical distance is everything, process is nothing. Yet the paragraph laid out by Tversky and Kahneman blurs this distinction entirely and encourages the reader to think that the two cognitive activities are the same, although they are in fact quite different.

Thus, we see that not only are the major fields isolated from one another, but also the field of judgment and decision making itself is divided into groups of researchers who fail to recognize their differences in theories and methods. This division, like the former, also impedes progress in understanding the effects of stress on judgment.

Psychological Origins of Correspondence Competence

We Homo sapiens come into this world equipped with a remarkable degree of competence in our judgments about the physical world around us. By the time we are four or five years old we have become remarkably accurate in our perception of the objects and events in the physical world. And all the evidence suggests that Homo sapiens—and many other species—had this capacity from the very beginning of our (and their) appearance on earth. However, if it can be said that the last thing fish will ever discover is the water, it can also be said that the fish's indifference to water is matched by our indifference to our remarkable perceptual abilities. For it has only been in the last three centuries that Homo sapiens has seriously begun to investigate human perception, only to be greeted with a deep, and continuing, puzzle. The puzzle is this: How does it happen that the *three*-dimensional world can be so accurately perceived, despite its projection on a *two*-dimensional retina?

It is as significant as it is obvious that no one has to be *taught* to make accurate judgments about the physical characteristics of the objects and events around us. We don't have to go to school to learn how to make accurate perceptual judgments of the size or shape of objects, despite their changing appearance on our retina. In fact, most people don't know that this happens. It is equally significant that many other animals are as competent in this regard as we are (although there are some species differences). These facts strongly indicate that evolution is responsible for this kind of competence, the kind that I am calling *correspondence competence*, because it means that our perceptual judgments frequently and accurately correspond to the objects in the natural world over a wide range of conditions. (Hold up a plate and ask someone what shape it is and they will say a circle. Move the plate around; hold it almost sideways and show this to someone; they will still say "a circle," no matter how you present it. You have just demonstrated "shape constancy." "Shape constancy" and all the other constancies, size, auditory, etc., have been demonstrated many times in rigorous empirical experiments.)

Correspondence competence with regard to sizes, colors, shapes, and other properties of the natural world is such a valuable attribute that it surely provided an evolutionary advantage for any species that possessed it over any species that did not. Curiously, however, few anthropologists seem to appreciate the importance of this critical capacity. In all the speculations about the origins of human competence, this striking ability is seldom, if ever, mentioned. And the fact that it is general across many species, including lower orders as well as the great apes, leads to the strong inference that correspondence competence with respect to the natural world is derived from evolution; it is in our genes.

Psychological Origins of Coherence Competence

In contrast to correspondence competence, coherence competence is not easy for Homo sapiens to achieve; we are not born with it. Rather, we are unique as a species in that we are born with the *potential for achieving coherence competence*. Apparently, no other species has this potential. Indeed, it is only recently that even we began to acquire coherence competence in a serious form, probably when we told stories to one another, stories that had a beginning, a middle, and an end; perhaps about the time we constructed, or at least shaped, tools; probably before we constructed alphabets. But by the time of Pythagorus, coherence competence had become securely established. Although demonstrated during early childhood, our potential for achieving coherence competence requires education in order to be realized, and some of us never do realize it. One must be taught how to do arithmetic and algebra, how to read, how to cope with a syllogism, how to calculate probabilities, detect the inconsistency in an argument, as well as how to become computer literate. Experience in an unlettered society will never bring about that form of competence, nor will an untutored existence in a lettered society, as so many untutored citizens must discover. And although our very ancient ancestors—the early hunters and gatherers—could go through life without ever having to achieve much in the way of coherence competence—there was very little to be coherent about—going through life today without that competence is increasingly difficult because there is so much that does demand coherence competence. Thus, in contrast to correspondence competence, coherence competence is *not* in our genes, although the potential for achieving it is, and that fact has considerable significance for how we make judgments—professional, moral, political—under stress, a matter I discuss later. First, however, I want to draw a distinction between two aspects of cognitive competence.

Two Aspects of Cognitive Competence

Cognitive competence takes two forms; first and foremost is *subject matter* competence (often called *domain* competence); the second form is *judgment and decision making* competence. These are two independent matters; gaining competence with regard to subject matter does not guarantee competence with regard to judgment and decision making. It is a necessary but not a sufficient condition. One can think of domain competence as a storehouse of knowledge, whereas judgment and decision making refers to the execution, or application, of that knowledge. The former is largely a function of learning, memory, and deduction, the latter more a matter of observation and inference.

This is an important distinction to grasp, for it is often mistakenly assumed that domain competence assures process competence, or competence in judgment. For example, the first priority in becoming a physicist is gaining domain competence; you must learn the subject matter. But gaining domain competence does not guarantee competence in judgment under uncertainty in physics; physicists with equal domain competence will vary widely in the competence of their judgments. Similarly, one cannot make competent judgments about medical problems without the medical training that produces domain competence. But medical training must *also* include training in the process of medical decision making—which includes knowledge of probability theory and judgment analysis—if competence in the medical decision making process is to be achieved. Too often it has been falsely assumed that domain competence is synonymous with competence in the decision making process. Not only does accepting that false assumption lead to a waste of domain competence, it also can be dangerous, for subject matter experts can make foolish judgments as a result of ignorance about the decision making process. Thus, domain competence can be wasted by incompetence in the process of medical decision making; a storehouse of knowledge can be thrown away if it is executed badly. Acceptance of that conclusion has resulted in the addition of training in decision making in medical schools and the creation of a Medical Decision Making Society. (For a particularly cogent example of the failure of domain expertise, see the description of the decision making at the launch of the *Challenger* in Hammond, 1996.)

Domain expertise differs in terms of whether coherence competence or correspondence competence is required. When coherence competence is required, the expert must demonstrate that he or she knows how a system works, and must be able to describe the functional relations among the variables in a system. Thus, for example, when a medical student describes a patient's symptoms to his or her teacher, the teacher will often respond by asking, "do you know the mechanism for that," thus testing the coherence of the student's domain knowledge with respect to a particular physiological or biochemical system. But if the student simply mentions a disease entity, the teacher may ask, "what are the signs and symptoms associated with that disease," thus testing the correspondence competence of the student's domain knowledge with that particular disease.

But the student's competence with respect to the *decision making process* may never be tested or examined—in contrast to the emphasis on domain knowledge. In the case of coherence competence, the student may be asked to combine various pieces of probabilistic information and may do so intuitively, which the student will certainly do if she or he has received no training in judgment and decision analysis. But if the instructor has received no training in the decision process, then his or her intuitions will be pitted against the students'. It may well be that both will be wrong; however, neither will find out until a formal analysis of the decision task is carried out. And, as has been demonstrated many times, the errors may well be very large indeed, perhaps greater than would have resulted from an error in domain knowledge. In short, errors that result from the use of intuition in the attempt to achieve coherence are a threat to cognitive competence. And, as we shall see, this will have considerable importance for judgments under stress.

A different situation arises when correspondence competence in the decision making process is required. Domain knowledge within this metatheory consists mainly of knowing *which* of the multiple fallible indicators to look for and act upon. Because correspondence competence demands only empirical accuracy and ignores coherence, domain knowledge provides only knowledge of which fallible indicators to rely on. Within the correspondence metatheory, competence in the second aspect of knowing, knowing how to *use* the information from various indicators, constitutes the execution process. Knowing how to use the indicators may take training. Popular belief is that it is most important to learn know how to *weight* the indicators, but, in fact, it is more important to know whether one's judgment should be a *linear function* of an indicator (the more of the indicator the higher the judgment, as in the higher the temperature the sicker the child) or a *curvilinear function* (the more of the indicator the more of the judgment, up to a point, and then the more of the indicator the *less* of the judgment, as in the higher the temperature of the coffee the better, up to a point, and then the higher the temperature, the worse—one will be burned!).

Implications for the Study of Stress

The distinction between correspondence competence and coherence competence is of great importance for the study of judgments under stress because judgments within each of the metatheories may be differentially affected by stressful conditions.

Differential Susceptibility

One set of stressful conditions may exert a distorting effect only on correspondence judgments, that is, empirical accuracy, while another set may exert a distorting effect only on coherence, that is, rationality. Yet another set of conditions may do both. Or it may be that it is only the *amount* of stress that matters; that is, stress from any condition may affect both types of cognition equally, and the greater the stress the more each is affected. Or it may be that stress *enhances* one form of cognition and *degrades* the other. In short, these two different metatheories of cognition may be differentially susceptible to stress. These questions have never been addressed, however, because we have lacked an overall theory of judgment and decision making, a deficiency I am trying to remedy in this book.

Although we lack empirical studies of the sort I have just indicated, the theoretical concepts are sufficiently general and understandable that it is possible to demonstrate their utility. For a simple but cogent example of how stress can differentially affect the demands for coherence versus correspondence competence, consider the airplane pilot in World War I. He had few instruments: an altimeter, a compass, and perhaps an airspeed indicator and a turn-and-bank indicator. If he could not see the horizon, if he were immersed in fog or clouds, he was in serious trouble. He might even find to his surprise—and consternation—that he had been flying upside down. Because of the lack of instrumentation, he did not seek coherence, he had no coherence problems to solve; he was wholly committed to a correspondence theory of truth because he had to rely entirely on his perceptual apparatus. Were he to encounter bad weather, engine problems, or structural failure, he would provide an example of a person committed to a correspondence theory of judgment under stress.

Now consider the modern airline pilot who must go to ground school for many months in order to learn to make coherent use of information from numerous instruments to become a highly skilled rational/analytical decision maker. Stress for this pilot will not come from a loss of correspondence competence (inability to see the horizon), it will come from a lack of coherence competence due to contradictions or confusion among his instruments and, possibly, his cognitive activities. And that is what we saw in the case of Captain Haynes; he was at one moment immersed in an environment that demanded coherence competence and ignored correspondence competence, and in the next moment the situation was entirely reversed. Once the hydraulics system of his airplane failed, his environment demanded correspondence competence which, to his credit and that of his colleagues, he was able to achieve (almost). We really don't know what the effects of this stressful situation were on Captain Haynes's cognitive activity, but his reports suggest that, in fact, they were minimal. We only know that our current knowledge would not have allowed us to predict the behavior that ensued, nor do we know how many pilots would have failed this test.

So we need to know the differential effects of stress on correspondence competence and coherence competence. And because no other psychologist in the judgment and decision field has taken this approach, I am confident that it will be new to the reader. Therefore, I have taken several steps to try to make these ideas comprehensible. It *is* regrettable that these distinctions were not introduced earlier, for the result is that we do not know whether it is rationality or accuracy that is more susceptible to stress, or whether different stressful conditions affect judgment differently. Does an evolutionary perspective shed light on the answer to this question? It does.

An Evolutionary Perspective

To see how an evolutionary perspective helps to differentiate cognitive responses to stress, recall that competence in correspondence judgments provides ecological fitness and thus evolutionary advantage. Coherence competence, or the achievement of rationality, on the other hand, must come through participation in a culture that educates its members in coherence competence. Coherence competence, when it appears, does so at the encouragement, assistance, and demands of a culture. Speaking broadly, then, correspondence competence is genetically determined, whereas coherence competence is culturally determined. (See Hammond, 1996, for a fuller account of this distinction.)

If that is so, then we should expect to find correspondence competence to be more resistant to stress than coherence competence. Why? Because correspondence competence "comes naturally" to us because of our genetic inheritance; we exercise our correspondence judgments almost instantaneously *without thinking,* just as we correctly perceive the world around us without thinking, without making strong, conscious demands on memory—an enormous advantage in the world of our ancestors subject to attack by a variety of predators. And that competence—

in visual, auditory processes as well as in other forms of perception—is resistant to stressors, or so it seems; we have no systematic, calibrated studies of this.

But coherence competence is different; it must be acquired, therefore it depends on the opportunity and ability to *learn* to engage in coherent cognition, to make coherent judgments, and on our ability to remember what we have been taught, as well as our ability to apply our training correctly. As a result, coherence competence requires *time*—time to learn and time to think when applying what we have learned. Thus, it is susceptible to time pressure. It is also susceptible to such distractions as noise, vibration, sleep loss, all of the stressors that are inimical to thinking, all of the stressors favored by stress researchers. In short, *correspondence competence will be more resistant to stressors than coherence competence.* And therein lies the significance of this distinction for the study of judgment under stress. The problem is: We don't know what the parameters of these relationships are, and, as I will show below, this hypothesis may be reversed, depending on what sort of circumstances are encountered.

Consider Captain Haynes once more. Suppose that he had not been deprived of all of his means of controlling the airplane, but that *some* of his controls didn't work as they should. That is, suppose he lost only *some* of his coherence competence (not an altogether unusual event!) What then? Would (does) calmness prevail? Is stress a linear function of the degree of loss of coherence competence? There are probably numerous examples of the occurrence of such situations but I know of no serious studies of them, although they would not be difficult to carry out in the simulator. The same must be true of operators of power plants and other installations that require coherence competence. In short, we don't have good studies of the effects of the stress that is endured when coherence competence is reduced in various degrees.

Although we know little about the differential effects of stress on people's judgments made within the two metatheories, we can predict the consequences of errors that are produced within them. Errors made within an analytical mode of cognition, within either metatheory, are very likely to be large, even catastrophic, whereas large errors are less likely to be made within an intuitive mode (see Brunswik, 1956; Hammond, Hamm, Grassia, & Pearson, 1987).

At this point, the reader need only observe that it is possible to separate the two metatheories in terms of their function in the judgment process, and that doing so provides insight into why some cognitive functions are more readily subject to stress and why some are not. Distractions vary in their character much more than I have indicated so far, however, and therefore it will be useful to distinguish among distractions, just as I have distinguished among metatheories of cognition, and it is to this topic I now turn.

Exogenous and Endogenous Disruptions

Exogenous disruptions are those that arise from outside the system—for example, fire, noise, heat, cold, interruptions from physical breakdowns *outside* the system. Endogenous disruptions are those that emerge from the task situation itself. Examples include a faulty gauge, an error-ridden report or reporter, time pressure, brakes that don't work, a computer program with bugs; these are disruptions or stressors that arise *within* the task system.

Differential Effects

Now we can see that the differential effects of stress on cognitive processes might well depend on the type of stressor—exogenous or endogenous. For we can anticipate that since correspondence competence is in our genes, and since as a species we have long experienced exogenous stressors, resistance to such stressors should be in our genes as well. Therefore, such stressors should have differential consequences for correspondence competence and coherence competence. We have no evolutionary history to help us when striving for coherence in our judgments, and therefore judgments based on coherence should be less resilient to these exogenous stressors. Of course, the reverse should be true. The reader will note that I have been tentative in my description of the differential effects of exogenous and endogenous stressors on correspondence and coherence competence. These statements are in fact hypotheses that require testing. And to be tested they need to be made more specific. I will be more specific in part II when I put forward a theory of judgment and stress. But first we need to see the place of intuition and analysis within each of the metatheories.

4

Correspondence Theories and Their Implications for Judgment Under Stress

I now briefly describe several research programs that focus their efforts on attempting to understand the correspondence between person's judgments and empirical events. I also indicate the potential usefulness of these research programs for the study of the effects of stress on judgment within the correspondence metatheory.

Signal Detection Theory

Signal Detection Theory (SDT) had its origins in Broadbent's (1957, 1958) theory of *attention*, in which he postulated that a sensory input must first pass through a selective filter and a channel of limited capacity before detection. Differences in the level of attention given to a stimulus could lead an observer to set the *decision criterion* higher or lower for determining whether a stimulus is present. That is, a less attentive person would set the decision criterion higher before saying the stimulus is present, a more attentive person would set the decision criterion lower. Broadbent's work led to a landmark book titled *Decision and Stress* in 1971. The reason the word "decision" appeared in the title was because Broadbent required his subjects to *decide* between the presence and absence of a stimulus. And that led to the theory of *signal detection* (Green & Swets, 1966/1974; Swets, 1973).

SDT partitions human judgment behavior into two independent components: an information processing component that builds internal representations of external events and a decision component that generates responses. The information processing component builds representations of two categorical events (e.g., whether a visual stimulus is present or absent, or whether a defendant is guilty or innocent), which are modeled as overlapping Gaussian probability distributions ("normal, or bell-shaped," curves). The accuracy with which the information processing component can separate the two distributions is a function of the size of the differences between the means and the magnitude of the standard deviations of the two distributions. In short, SDT is a correspondence theory. That is because it is concerned entirely with the empirical accuracy of the person's decisions; nothing is said about the coherence or *rationality* of the processes that produce the decisions.

For example, a simple decision rule might be: State that the event occurred if the evidence exceeds the criterion, or state that it didn't occur if the evidence does not exceed the criterion. There are four possible outcomes for each decision, the relative frequency of which corresponds to different areas under the event probability distributions bounded by the decision criterion. Only two of these outcomes are independent: the hit rate (the probability of predicting an event given that the event occurs) and the false alarm rate (the probability of predicting an event given that the event does not occur, also called *false positives*). The other possible outcomes are a correct rejection (the probability of not predicting an event, given that it does not occur, which is the complement of the hit rate) and a miss (the probability of not predicting an event, given that it does occur, which is the complement of the false alarm rate, and is called a false negative). The relative probabilities of these four outcomes are a function of both the judge's sensitivity to the information and his or her choice of the location of a decision criterion.

There are two things that make SDT very useful. One is its ability to differentiate the four aspects of empirical accuracy (the hit rate, the correct rejection rate, and false positives and false negatives). Thus, it becomes possible not only to examine past decision making behavior in terms of hits and correct rejections, but also to adjust decisions to meet desired specifications, such as minimizing false positives. Thus, SDT's specification of a decision criterion that can be shifted to change the relationship among these four categories is also very useful for examining performance with respect to professional, moral, and political judgments.

SDT has not, to the best of my knowledge, been used to study the effects of emotion on judgment and decision making, although it certainly could be, and should be. The list of emotions set forth by Lazarus and Lazarus (1994) makes it easy to see how different emotions might well lead to different errors of judgment, or possibly, to differences in accuracy of judgment. For example, what they label as the "nasty emotions: anger, envy, jealousy" suggest that a person under the influence of these emotions might make errors, or decide on a different allocation of risk to different parties than they would had they not been under their influence. Or these emotions might lead persons to demand increased, possibly unattainable, performance from a subordinate, or a harsh allocation of risk might be induced by these "nasty" emotions. On the other hand, what they call "existential emotions: anxiety-fright, guilt, shame" might lead to the opposite, forgiveness, say, or a relaxation of standards. The same could be said for the other categories of emotions these authors list—"emotions provoked by unfavorable life conditions: relief, hope, sadness, and depression"; "emotions provoked by favorable life conditions: happiness, pride, and love"; as well as the "empathic emotions: gratitude, compassion, and those aroused by aesthetic experiences" (see their Table of Contents, p. xi). It is reasonable to suppose that any and all of these would have an effect on the location of the decision criterion that specifies the number of accurate judgments and the different types of errors that will be made under a given degree of uncertainty. If it were possible to induce the specific emotion of interest, and only that specific emotion, it would be a straightforward matter to ascertain its effect on the location of the decision criterion (over a variety of tasks). Some of the studies described next do just that.

Significance for the Study of Stress

A wide variety of important human judgment tasks can be studied within the SDT framework. Because SDT separates the effects of the observer's sensitivity to the data being presented from the location of the decision criterion, it makes it possible to ascertain whether a stressor independently influences one or the other or both. For example, a weather forecaster under time pressure might become either more or less sensitive to differences in the data being evaluated while steadily maintaining a specific decision criterion for announcing a critical weather situation. However, time pressure, or other stressors, may cause a forecaster to *shift* the decision criterion, moving it higher or lower, and thus affecting the likelihood of false alarms and/or misses, although his sensitivity to the data (accuracy) remained the same.

A recent study applied SDT to just that problem (Harvey, Hammond, Lusk, & Mross, 1992; Lusk, 1993). These researchers were studying forecasters who were predicting whether severe weather (thunderstorms) was about to occur at approaches to an international airport. Increased stress in the form of time pressure was found to *increase* the forecasters' accuracy. At the same time, however, they were shifting their decision criterion so that they issued fewer false alarms (fewer false positives). Thus, it was ascertained that time pressure induced the forecasters not only to become more accurate but also to change their policy for allocating risk. Under the stress of time pressure it became more important for them to decrease the error of issuing false warnings of dangerous weather (that is, decrease the likelihood of unnecessarily rerouting and delaying air traffic) and to increase the risk of not issuing a warning when they should have (thereby increasing the risk of a weather-related accident). Use of SDT thus documented a change in *both* accuracy of forecast (it increased) and allocation of risk (reductions in unnecessary rerouting). Thus, SDT provides a nice example of the separate *conceptual* denotation of fact (accuracy) and value (allocation of risk). SDT shows their inseparable *empirical* relationship under uncertainty; one error can't be decreased (increased) without the other being increased (decreased) concomitantly, as long as the accuracy of prediction remains unchanged (see Lusk, 1993).

Examination of the errors of polygraphers attempting to discriminate the justly from the unjustly accused have been infrequent, but K. Hammond, L. Harvey, and R. Hastie (1992) have shown that some polygraphers (and their clients) are unaware of their choice of risking false positives or false negatives. Medical doctors making diagnoses based on medical imaging techniques such as computed tomography and forecasters trying to predict tomorrow's weather (e.g., Lusk, 1993) are also unaware of their performance regarding the choice of errors. Indeed, many professionals are completely unaware of the possibility of learning about their performance.

In an effort to examine the effects of a shifting decision criterion in a dynamic task, that is, one in which conditions are changing, J. Kerstholt (1995) systematically manipulated the probability of false alarms. Examining decision making behavior under changing conditions of uncertainty outside the laboratory is difficult and rarely done (but see Lusk, 1993). Kerstholt wanted to know what would happen to decision making behavior as the likelihood of false alarms changed from low to high, and he was particularly interested in the effects of time pressure on decision making. He found that "the more subjects expected a false alarm, the longer they waited before making information requests or taking actions. In contrast, time pressure did not affect the relative time after which subjects started to intervene" (1995, p. 193). The graph in Kerstholt's Exhibit 5 (p. 193) illustrates the relative effect of false alarm rates (large) and that of time pressure (small). Kerstholt also found that different false alarm conditions did not affect the time it took to process the information, that is, to decide what to do. Time pressure, not surprisingly, did result in more rapid processing behavior (p. 194).

This single study cannot of course provide us with definitive conclusions about the effects of differing false alarm rates and or time pressure on decision making behavior; it does, however, have the important result of showing us that such behavior can be usefully studied under dynamic task conditions. Agencies that require their employees to face such conditions should learn that and undertake the sponsorship of research programs that will enlighten their professional work.

Assets and Limitations of SDT

Thus, after a long history of laboratory research, SDT has demonstrated, in both theoretical and empirical terms, its utility for ascertaining the effects of stress on judgments under uncertainty, and its general value in many different field conditions. Those who argue for the potential value of laboratory research for the social good it might produce will find much comfort in the history of SDT; it is a classic case, and that is why John Swets was awarded honors. Critics of the value of laboratory research, on the other hand, will have to acknowledge that this is an "exception" that embarrasses their point of view or find an explanation for how it happened to be such a success. For my part, I choose the latter course.

My explanation is that the successful generalization of the applicability of SDT from laboratory to field is due to the fact that it encompasses the *formal properties* of tasks involving irreducible uncertainty. That is, SDT describes task circumstances in *formal* terms (the relative frequencies of events) and judgments made in response to those events. As a result, it is *domain independent*. That is, SDT does not include the *substance* of these events (they can be thunderstorms or any other environmental events). SDT requires only that the statistical properties of the events be described. And that is exactly why the next theory (Social Judgment Theory) to be discussed also provides generalization from laboratory to field; it too focuses on the formal rather than the substantive properties of both task and subject.

SDT does have its limitations, however. For although SDT describes the policymaker's behavior in terms of sensitivity and the location of the decision criterion, it does not offer descriptions of the cognitive activity that produces the judgment in response to the events observed. And although it identifies and describes the values that are explicated in the choice of the location of the decision criterion, that is, how many false positives and false negatives are to be tolerated, SDT does not explain these choices in terms of the observer's judgment policy. And it is that policy that produces the hits, misses, and the false positives and false negatives. The next research program directs itself to that problem.

Social Judgment Theory

Social Judgment Theory (SJT) grew out of the theoretical and methodological approach put forward by E. Brunswik (1943, 1952, 1956; see also Hammond, 1966, 1996), in which this conceptual framework is described. Brunswik's (1955) theory of cognition embraced uncertainty as characteristic of both the environment and the organism. His approach is based on the premise that higher organisms (and many lower ones) receive information from the environment in the form of multiple fallible indicators of (or "cues" to) some unobservable state of the environment (where the food is likely to be, who is an enemy, which is the appropriate sex partner). Brunswik's approach thus falls within the correspondence metatheory because it presupposes that the organism intends to be as empirically accurate as possible in its judgments about environmental objects and events. Empirical accuracy clearly contributes to ecological fitness in a natural environment, and therefore we may suppose that it would be a basis for natural selection. Although originally developed for the psychology of perception, the Brunswikian framework was adapted by Hammond (1955 et seq.), Brehmer, Doherty, Stewart, and others to the topic of judgment (for an overview, see Brehmer & Joyce, 1988; for an early example, see Hammond, 1955; for further explanation, see Cooksey, 1996; Hammond, 1996).

Lens Model Diagram

The lens model (see figure 4.1) was introduced by Brunswik in 1934 (see also 1956) to provide a pictorial representation of the multiple fallible indicators—the mediation of information *from* an environmental state—as well as the *use* of these multiple indicators by the organism. Note that the lens model describes only the formal properties of both task and subject, and thus, like SDT, is a *domain independent* theory.

Multiple Fallible Indicators

The concept of multiple fallible indicators follows from Brunswik's (1934, 1943, 1956) suggestion that perception is based on "probabilistic cues." (These are portrayed in the lens model diagram as X_is.) The idea is simple and familiar; objects in the environment, including people, exhibit various *palpable* attributes, attributes that one can see, feel, or smell, that is, attributes that are available to our sensory capacities. It is these

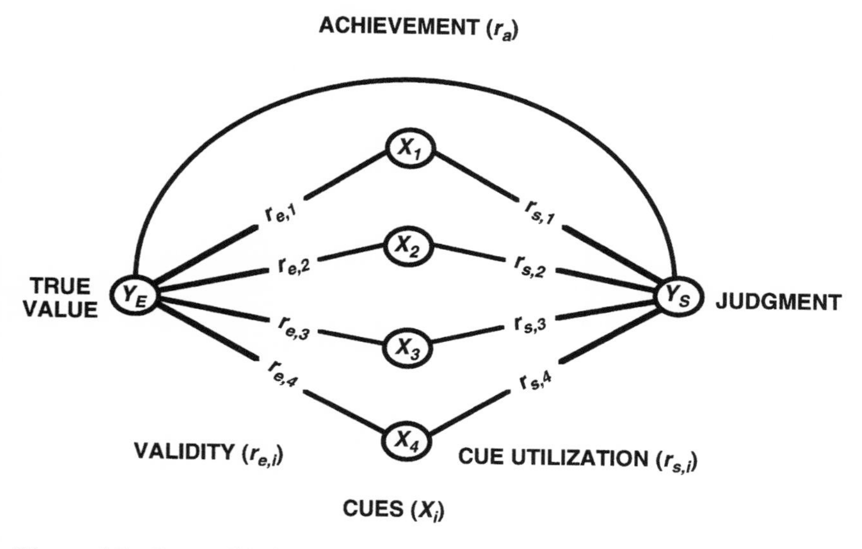

Figure 4.1. Brunswik's lens model.

palpable attributes that function as indicators of various *impalpable* aspects of the objects. Put otherwise, cues, or indicators, are those features of the objects that you can make use of to infer those aspects of the objects that are not directly available to you.

Our social life is made up of that cuing activity, inferring the deeper characteristics of another person. For example, because you cannot directly see another person's intentions toward you, they must be inferred from what you *can* see, namely, his or her facial expressions, actions, and activities (often called "body language") as well as from the form and content of their language. Of course, you must infer many other aspects of another individual's *persona*—their honesty, trustworthiness, intelligence, political beliefs, moral value systems—indeed, all those characteristics that will determine your relationship with that person. All those inferences are based on multiple fallible indicators; *multiple* because there are so many of them, and *fallible* because none is perfectly dependable, or accurate, in their ability to indicate that deeper characteristic that you are trying to infer.

Not only are multiple fallible indicators used in developing our interpersonal relations, but they are also central to our physical orientation, the perception of distance, and the size and shape of objects, and so forth. And we *construct* indicators when our environment does not provide them; we do that because it is in our nature to make use of multiple fallible indicators. Thus, we construct all those indicators of the public health, financial markets—the Index of Leading Economic Indicators is the best known—and others. And when it is critical to make accurate distinctions among us in terms of rank, for example, we construct indicators to be *infallible*. Thus, for example, we use insignia (one, two, or three chevrons in the army, or one, two, or three bands on the sleeves of naval officers) to make it perfectly clear who is subordinate to whom. And when it is important to have infallible indicators in other circumstances—road signs, for example—we construct these as well. Nor are we the only creatures who make use of multiple fallible indicators; every bird or animal that navigates makes use of the stars, temperature, wave direction, or other fallible indicators to find its way to its breeding or feeding grounds (see Hammond, 1996, for further development of this concept).

Lens Model Equation

A significant development in SJT occurred with the introduction of the lens model equation (LME) by Hursch, Hammond, and Hursch (1964). It is this equation that provides a set of parameters that can be used not only to delineate the parameters of accuracy, but to ascertain the effects of stress as well. The lens model equation generally takes the following form developed by L. Tucker (1964):

$$r_a = GR_eR_s + C\sqrt{1-R_e^2}\sqrt{1-R_s^2}$$

where

r_a represents achievement, that is, the correlation between a subject's judgment and the criterion of accuracy for that judgment;

G represents achievement when the linearly predictable uncertainty in the environmental system and the subject's response system is removed;

R_e equals the linearly predictable variance in the environment; and

R_s equals the linearly predictable variance in the subject;

C equals the correlation between the residual variance in the environmental system and the subject's prediction system, and thus represents the correlation between nonlinear components of each system.

The general framework of SJT, the lens model, and the lens model equation have been applied to a wide variety of problems, including learning under uncertainty; interpersonal learning (Gillis, Bernieri, & Wooten, 1995); conflict resolution (Hammond & Grassia, 1985), weather forecasting (Stewart & Lusk, 1994); child abuse (Dalgleish & Drew, 1989); and the effects of psychoactive drugs (Hammond & Joyce, 1975); among others (see Brehmer & Joyce, 1988, for a review; see also Hammond, 1996, for examples of the application of SJT to the topic of social policy formation; see Cooksey, 1996, for a definitive and detailed explanation of the lens model equation and its applications to a wide variety of problems). It is important to note that SJT is not restricted to being modeled by the statistics of linear regression, as has been mistakenly claimed on occasion.

Generality

SJT parallels SDT in claiming that both theory and method generate results from laboratory research that also apply to conditions outside the laboratory, and it does so for the same reason; both theories describe task conditions in terms of their formal properties. As noted above, SDT provides for the categorical description of uncertain objective environmental events *of any kind*. That is what is signified by the term *formal;* it implies that the theory does not contain any specific content, such as thunderstorms or special laboratory "stimuli," and thus is devoid of *substantive* properties. And because of their mathematical structure both theories achieve coherence, although both pursue the study of that form of human judgment that aims at correspondence, that is, accurate prediction of events in the environment.

SJT also provides formal terms for the denotation of objective task conditions. Specifically, R_e specifies the amount of uncertainty (or unpredictability) in the task, and r_e specifies the ecological validity of an indicator of (or cue to) the object or event to be inferred or predicted; both can be measured, irrespective of the substantive nature of the materials involved. SJT also specifies the linearity or *nonlinearity* of the relations between a cue or indicator and the object or event to be inferred or predicted. In short, theories such as SDT and SJT that are formal in nature, and thus specify the formal properties of the task and the properties of the human system as well as their relationships, are more likely to establish the generality sought by all researchers than theories that do not.

Significance for the Study of Stress

The parameters of the LME can readily be tested for sensitivity to stress, or more precisely, to changes in task conditions. Rothstein (1986), for example, examined the effects of time pressure and found that achievement (r_a) decreased in a task involving judgments under uncertainty. He discovered that the decrease in performance was due to a decreased R_s (consistency in the subject's *execution* of his/her judgments) rather than to a *deformation* in the subject's knowledge (G). That is, the subject continued to use the information from the multiple fallible indicators in the appropriate manner; it was the consistency of execution that changed. If Rothstein's results are generally true, they carry significance, for they show the utility of the LME for distinguishing between the impact of disruption on *knowledge* (G) or *control* (R_s) of the execution of those judgments (see Cooksey, 1996; Hammond & Summers, 1972). (This result parallels one obtained by the use of Signal Detection Theory by Hammond, Harvey, & Hastie, 1992.) And since it will be the case that some stressors (noise, for example) will have no effect on the objective uncertainty in the task, whereas in other cases stress will be evoked because of a breakdown in task conditions (system failure) and thus task uncertainty will be increased, it is essential that this distinction be central in the investigator's mind. If not, there is a danger that it will be wrongly concluded that the *person* rather than the *task* environment had changed. Of course, the performance of the person may deteriorate (become less consistent), but the change in the task conditions must be examined to see whether they have increased in uncertainty, and thus induced a decrease in the consistency of the subject's judgments.

An important and often overlooked feature of the lens model equation is that it allows the examination of nonlinearity in judgment tasks and judgments and the relation between the two. That is important because it makes it possible to examine and quantify the effects of stress on nonlinearity, for example, a curvilinear relation between a component of the task and the judgment system. And because the C term in the equation represents the correlation between the nonlinear component(s) in the task system and the nonlinear components in the subject's judgment system, it makes it possible to quantify exactly those relations. Curvilinear relations between cues and judgments should be harder for a person to maintain under difficult circumstances than linear ones, but there has been no research on this topic (see Cooksey, 1996, for a detailed discussion of the LME).

In addition to separating a change in execution (R_s) from a change in knowledge (G) in performance (r_a), the LME also suggests to the investigator that the task situation needs to be described both before and after the stressful events have occurred. In the case of linear relationships, if the uncertainty in the task has increased (thus decreasing R_e) because of a disruption, then the upper limit of accuracy will be reduced, inasmuch as r_a cannot exceed R_e. This limit is of considerable importance in the evaluation of performance (r_a). For when the task conditions are changed by disruption, the uncertainty in the task is ordinarily increased.

Any theory of task conditions immediately brings to our attention the fact that task conditions can be changed in many ways, and it is the duty of the theory (or theorist) to specify exactly what those changes might be, as well as what they are at any given time. And because stress almost always implies change, it becomes necessary to indicate which theory-specified parameters of the task have changed. The theory must then indicate what the behavioral consequences of that change in task parameters will be. The lens model equation of SJT allows us to do that, as the following example will indicate.

Numerous analyses have been made of the shootdown of the Iranian airliner by the USS *Vincennes* (see chapter 1), but there is general agreement that the scene in the control room of the ship was one of confusion and uncertainty (e.g., see Zsambok & Klein, 1997). It can also be described as one in which the controlling officer was responding to a situation of irreducible uncertainty, at the time the decision to fire on the target was made. And it can also be described as one in which the officer was responding to multiple fallible indicators of the general situation (e.g., the location and speed of the target, its direction, its lack of communication and events on the surface that might indicate an attack on the *Vincennes*). That situation can be visualized in terms of the lens model and its associated theory of tasks. For the officer in charge, the ecological validities of the indicators were unknown; that is, he could not be sure that he was getting accurate information about any of these indicators, and it turned out that in fact he was not. As a result, R_e decreased remarkably, which probably reduced R_s. This meant that because his uncertainty about the information he was receiving—and how to organize it—increased considerably, his response (his judgment) became essentially unpredictable.

Consider a second example that I shall describe in abstract terms. Suppose that a commander of a fire fighting group must make a judgment about the nature of the fire he must extinguish. His judgment will be made on the basis of multiple fallible indicators such as the height of the flames, the color of the smoke, information about what the building contains, etc. Now suppose that some unanticipated event occurs—the roof falls in—that immediately requires a new plan of action. This is a stressful event. Will the commander's focus of attention narrow? Will he or she focus on only one or two indicators of what is happening? And if so, will that lead to poor judgment? That is what the experts testifying before the House Armed Services Committee said would happen. That is what some textbook writers tell their readers will happen. Indeed, it is the conventional wisdom. But the conventional wisdom is called into question by current researchers (see Wickens, 1996) who find studies in which this does not occur. Two investigators, Gigerenzer and Goldstein (1996), go further, however. They suggest that "narrowing" may actually *improve* the accuracy of judgments. Could this actually occur? I will take that question up in the next section.

Probabilistic Mental Models

Gigerenzer's primary goal has been to show that the conclusions drawn by the coherence theorists, Tversky and Kahneman, about the incompetence of human judgment are incorrect because of their use of flawed methods and lack of a theory. The model of judgment that he has put forward is similar to the lens model in that it is a correspondence theory that places great emphasis on the concepts of ecological cue validity and reliability. Therefore, I turn directly to the implications of Gigerenzer's most recent work for the study of stress.

Significance for the Study of Stress

Gigerenzer and Goldstein (1996) begin by presuming, as does the lens model, that a person normally makes use of multiple fallible indicators in making judgments under uncertainty about some presently intangible event or condition. They found, however, that subjects who used the *single* most ecologically valid indicator were the most accurate, a strategy Gigerenzer and Goldstein termed "take the best." And they substantiated this empirical finding with a mathematical analysis that indicated this is what a person seeking empirical accuracy (correspondence) *should* do. That conclusion directly challenges the folk wisdom that "narrowing" the focus of attention—*if* stress does that—leads to poor, that is, inaccurate judgment, and perhaps disaster. Indeed, work by Gigerenzer and Goldstein suggests just the opposite of what Zornetzer and other experts presented before the members of the House Armed Services Committee in the 1989 hearings (see chapter 1). In short, even if "narrowing" one's scope of information does occur (and there are strong doubts that it does), a narrower focus does *not* necessarily reduce the accuracy of judgment; indeed, it may improve it. If that is true, it is an important research finding.

The reader should note an important difference between Gigerenzer and Goldstein's "take the best" approach and that of SDT and SJT. Gigerenzer and Goldstein's model is "memory-driven," in that it requires the subject to search his or her memory for the various cues to be used in their judgments; the SDT and SJT models, however, are "information-driven." That is, the subjects have all the information they need in front of them. This distinction leads to some important qualifications to the Gigerenzer and Goldstein findings in relation to stress.

The first is that the person making the judgment must *know in advance*—and remember—which of the several cues, or indicators, *is* the best, in order to "take" it, and that information may be unavailable. Ask yourself which of the cues in the hierarchy of perceptual cues to the intelligence of another person is the best. Do you know? Does anyone know? Or ask yourself which of the several perceptual cues to dangerousness of another person is the best. If you don't know, you have lots of company. (And you may have to make such judgments under time pressure.) There is, in short, no guarantee that people will *know*—or remember—which indicator in the hierarchy is the "best" (has the highest ecological validity). Therefore, there is no guarantee that they will automatically *choose* the best, most ecologically valid, cue. Moreover, if there are a large number of cues, and if the ecological validity of the cues is roughly equal, then it won't matter much if the "next-best" cue is chosen, rather than the "best."

A second qualification to be made with the "take the best" strategy is that one must consider the intercorrelations among the indicators. If there is a zero correlation between two indicators with equally high ecological validity, it means that each indicator has a separate contribution to make to accurate predictions of the criterion; Predictor A predicts one aspect of the criterion, Predictor B predicts another, independent, aspect. Therefore, taking account of both, that is, depending on both for a prediction, will result in a more accurate prediction than relying on either one alone. This will also be true if one predictor is only somewhat better than the second; adding the information from the *second-best* predictor (provided it offers independent information) will improve the accuracy of the prediction. But in many if not most judgment situations, particularly those that demand intuitive judgments, predictors will not provide independent information, but will be redundant with one another. And to the extent that they are redundant, it will matter little which predictor is used, and there will be little to gain by adding information to that provided by the best predictor. Therefore, in that case the "take the best" strategy will in fact be the best, or at least, as good as any other.

In short, the idea that narrowing the focus of attention is necessarily detrimental to judgment is certainly not one that can be applied to all situations. And that may be why researchers have reached different conclusions about this matter; they have studied noncomparable situations. The solution to this problem lies in the development of a *theory of task conditions* that will allow predictions of which circumstances will and will not produce narrowing and with what consequences. I discuss this matter in more detail in later chapters.

5

Coherence Theories and Their Implications for Judgment Under Stress

Coherence is a psychologically compelling attribute of any argument, but it is also a *logically* compelling attribute; indeed, it is a mandatory attribute for those arguments—or judgments—that do not rely on their correspondence with empirical facts. Consider a jury trial in which the defendant's attorney offers an alibi for the defendant; he was at a certain place at a certain time that made it impossible for the defendant to have committed the crime; therefore, the attorney claims his client is innocent. What will the prosecutor's strategy be? It will be to show that the opponent's argument lacks consistency; that it contains a contradiction; that it places the defendant at two different places at once. In short, it is incoherent, therefore it can't be true.

Mathematics provides the ultimate test of the coherence of any argument because mathematics prohibits contradictions. And it is probably for that reason that an argument is often criticized in colloquial terms by simply saying that it doesn't "add up." And so it is in the sciences in general; formulate your argument (or theory) in mathematical terms in a series of interlocking equations and deductions therefrom, and although your theory may turn out to be empirically false, you will at least have established some credibility for having presented a coherent theory. Albert Einstein, for example, famously formulated his theory of relativity in purely mathematical, and thus highly coherent, terms. Although the world waited anxiously for the results of an empirical test of his theory, Einstein claimed he had no interest in the test because his mathematical proof had convinced him of the truth of his theory. Fortunately, he was right. Kurt Lewin, a prominent psychologist in the post-World War II period, made a similar attempt to develop a purely mathematical theory for psychology, but his work is long forgotten because no empirical proof was ever forthcoming.

The field of judgment and decision making has found a different use for the coherence offered by mathematical statistics, namely, a *standard* for evaluating how people reason when making judgments under uncertainty. Borrowing such concepts as sampling, conditional probability, and most of all, Bayes' theorem from the field of mathematical statistics made it possible for psychologists studying the rationality of human decision making to establish a mathematical standard for evaluating the coherence, and thus the *rationality and competence,* of decision making under uncertainty. That was no small step. For once a standard of such impeccable character became available, arguments about *cognitive competence* could be settled, and the correct answer to many problems could be found. Moreover, competence could be expanded as knowledge about mathematical statistics and probability was expanded.

Psychologists carried out studies of the cognitive competence of people by asking them to make decisions based on the same information that was given to an equation derived from mathematical statistics and then comparing the answers. Should the answers differ, it would be concluded that people were irrational in their decision making. For if their intuitive judgments did not conform to the analytical answers produced by the mathematical equations, their decision making processes would have to be considered to be flawed, that is, incoherent. Ward Edwards was one of the first psychologists to carry out such studies in a systematic way under the rubric of "decision analysis" (Edwards, 1954).

Decision Analysis

The decision analysis approach to judgment and decision making began in the 1950s and 1960s when psychologists discovered the theories of optimal behavior proposed by mathematicians, economists, and statisticians (e.g., Raiffa, 1968). One such theory is the theory of expected utility (EU) developed by von Neumann and Morgenstern (1947). They demonstrated mathematically that, if a decision maker's choices follow a certain set of axioms, it is possible to derive expected utilities for each choice (the sum of the expected value of each potential consequence of the choice multiplied by its probability) such that a given choice will be preferred to an alternate choice only if its expected utility is higher. EU theory therefore describes a "normative" or rational model of how a decision maker should combine expected utilities to make optimal choices.

Another influential normative theory is Bayes' theorem, which was proposed by Savage (1954; see also Edwards, 1954) as an optimal mathematical formula for using new data to update prior beliefs about the probability of an uncertain event. Recognizing that these theories represented *formal models of human judgment*, psychologists began to test them empirically in the laboratory and to suggest revisions. Edwards (1954), for example, demonstrated through a series of experiments on risky choice that EU theory must take into account the *subjective* nature of probabilities as well as their objective, numerical values.

There are two basic ways in which decision analysis models can be used. First, they afford a *prescriptive* analysis of the judgment process in terms of a normative theory (e.g., SEU or Subjective Expected Utility theory) of how decision makers *should* behave, if they are to claim rationality. That is, they provide a standard of rationality. Second, such models can provide a set of parameters that can be used to make a *descriptive* analysis of human beliefs and values and the manner in which they *are* incorporated into actual judgments. Subjects' judgments may then be compared with those produced by the normative theory; and the comparison thus forms a "yardstick" for determining how close human judgment behavior is to the best it could possibly be. The decision analyst focuses on comparing the rational model with measured behavior.

Measured judgments can deviate from the normative models in one of two ways: (1) by being less *consistent* (i.e., following the axioms of the normative model but erratically) and (2) by being less *coherent* (violating the underlying assumptions of the normative model). Many studies have reported probabilistic inferences to be less coherent than they should be, violating the assumptions of the Bayesian model, for example, by failing to take into account relevant base rate information (Kahneman & Tversky, 1973; Plous, 1993; but see Koehler, 1996, who claims this error occurs far less frequently than has been claimed).

When people deviate from the rational model, the typical response of decision analysts is to use decision aids to help them implement the rational model. Raiffa (1968, 1969), for example, developed Multiattribute Utility Theory to extend SEU theory to the case when each choice option has multiple attributes, and it has been used to aid judgments in a variety of situations (see von Winterfeldt & Edwards, 1986, for a description of the use of decision aids; see also Keeney & Raiffa, 1976).

Significance for the Study of Stress

The major advantage offered by decision analysis for studies of stress and decision making is that it provides a standard for assessing coherence. If one knew how far people ordinarily deviate from Bayes' theorem under benign circumstances, then one could determine the effects of stress; for example, does stress induce persons to deviate even further from rationality? It is possible, for example, that stress may affect the use or neglect of base rate information (see above) in making their judgments. Overall, decision analysis makes it possible to ascertain whether stress affects either (1) the decision to adopt a normative versus some other model or (2) the values of the subjective parameters (e.g., subjective probability or decision weight) if a normative model is adopted. However, at present there is no theory that can predict which parameters would be expected to change in which direction under stress. Decision analysis can, however, provide prescriptions for improving decision making under stress in the form of decision aids based on rational models.

The "log odds" form of Bayes' theorem (log posterior odds = log prior odds + log likelihood ratio) would prove especially useful for ascertaining the effects of stressful conditions because of its simplicity and its psychological significance. The "log posterior odds" part of the formula corresponds to, and can be compared with, the judgment of the subject regarding the likelihood of an event. The "prior odds" part refers to the "base rate" occurrence of the event. The "likelihood ratio" refers to whatever other evidence may be available to the subject (or investigator). Thus, if the investigator wishes to ascertain the relative impact of base rates versus other information on the subjects' judgment, this formulation makes it easy to do so. And, indeed, various researchers have investigated this question (see, e.g., Kahneman & Tversky, 1982), leading to the conclusion that people frequently "ignore" base rates in favor of other evidence.

The question of interest is whether stressful conditions lead to changes in either component of the above equation; that is, does stress lead to *better* or *worse* subjective judgments of the posterior odds of the occurrence of an event? And does either improvement or decrement in accuracy occur because of change in one component or the other?

Heuristics and Biases

The approach to human judgment known as "heuristics and biases" arose out of several separate lines of psychological research in the 1950s and 1960s which collectively demonstrated that human judgment is often less coherent than mathematical models such as Bayes' theorem demands. By 1954, Edwards was documenting substantial discrepancies between human inference and optimal Bayesian models. Simon (1955, 1956) was developing his theory of "bounded rationality," which postulates that individuals do not search for optimal choices but instead, due to time constraints and limited computational capacity, seek a solution which satisfactorily meets their level of aspiration and then terminate their search. Eventually, so many examples of judgmental error or bias had been collected that most researchers felt that no matter how useful normative models such as SEU theory and Bayes' theorem were as *prescriptive* models, it was time to abandon them for *descriptive* purposes and to search for alternate models of human behavior.

Judgment researchers turned to the emerging field of cognitive psychology for these new models, inspired by cognitive process models which emphasized limitations in such mental capacities as short-term memory (Miller, 1956) and attention (Broadbent, 1957), and how these limitations shaped more complex cognitive processes such as reasoning (Bruner, Goodnow, & Austin, 1956) and problem solving (Newell & Simon, 1972). Tversky and Kahneman (1974; see also Kahneman, Slovic, & Tversky, 1982), for example, proposed several simple cognitive processes (they called them "heuristics"), including representativeness, availability, and anchoring and adjustment (about which more below), which they claim underlie a broad range of human judgments. The advantage of such *heuristics* is that they reduce the complexity of judgmental tasks and make them more tractable for decision makers with limited mental resources, and often, it is claimed, yield answers that approximate the correct judgment. The disadvantage of these heuristics is that they introduce *bias* into the person's judgments and thus lead to error.

Researchers in the heuristics and biases tradition are, like decision analysts, interested in comparing the "judgments" produced by ***normative models*** of decision making with empirically derived ***human judgments.*** The goal of many studies in this tradition is to specifically search for situations that will induce subjects to produce judgments that are biased with respect to the normative model. That is, they seek out potentially error-producing tasks. (Kahneman and Tversky claim that this is in the tradition of psychological research—they are right—but many, including me, believe this is a *bad* tradition [Brunswik, 1956; Hammond, 1996].) Once such an error has been found, it is typically explained or demonstrated that "people" employ one of the three heuristics mentioned above, and it is the use of that heuristic that caused the error. It is proposed that although these heuristics often lead to approximately correct answers, they can lead to severe and systematic errors. It is the heuristic cognitive model, rather than the normative theory, that is used for subsequent description of the cognitive process.

Because normative theories are deemed to be inadequate for descriptive purposes, heuristics and biases researchers propose alternate descriptive models that incorporate heuristic rules. Kahneman and Tversky (1979), for example, have developed prospect theory as an alternative to SEU theory. Prospect theory extends SEU theory by incorporating an initial "editing" phase, in which various heuristic rules are used to simplify decisions by postulating that values are assigned, not to overall assets, but to gains or losses with respect to a reference point, and by assuming a generally steeper value function for losses than for gains. The theory can account for a variety of empirical effects which SEU theory cannot, including the tendency to overweight certain outcomes relative to others with the same expected value, and reversals of preference due to framing effects.

Another example of a formalized heuristic theory is the contingent weighting model proposed by Tversky, Sattath, and Slovic (1988) to explain choice between two options. In their model the weight or importance assigned to different input attributes of the choices depends on the nature of the output which is required. The theory can therefore account for response mode effects (e.g., reversals of preference depending on whether the decision maker is asked to choose between two options or to adjust the inputs of the two options so that they are equally matched) which Multiattribute Utility Theory (MAUT) cannot. Many of the proposed cognitive heuristics are not so well formalized as prospect theory and contingent weighting theory. The heuristics and biases replacement for Bayes' theorem, for example, turns to a collection or "toolbox" of heuristics that are often not well defined. Instead of using base rate information to judge probabilities, as Bayes' theorem requires, people may make their judgments by using representativeness (Kahneman & Tversky, 1972), that is, how similar the judged instance is to the relevant category. Or people may use the information that Bayes' theorem suggests, but use it in a different way, for example, by employing an "anchoring and adjustment" heuristic (Slovic & Lichtenstein, 1971), in which they overly "anchor" on one piece of information and adjust it, typically insufficiently, to take other information into account. However, heuristics and biases models at present provide no way of predicting which heuristic will be taken out of the toolbox and applied by decision makers in which situation.

Although heuristics and biases researchers defend their focus on errors and biases as sound research strategy rather than a deliberately pessimistic view of human judgment (e.g., Kahneman, 1991), the list of heuristics and biases is now so long (see, e.g., Kahneman et al., 1982)—and ever expanding—that the field is generally perceived as providing gloomy conclusions for human decision making ability. According to this view, the *coherence competence* of human decision makers is typically very poor, since people employ heuristics that deviate radically from the procedures prescribed by normative models. Lacking a single rational underlying strategy which they can apply in a variety of situations, people are very susceptible to *framing* effects (Tversky & Kahneman, 1981), and their judgments and choices are easily changed by minor differences in the surface structures of problems which are formally identical. Nevertheless, heuristics and biases researchers claim that heuristics often produce good answers; that is, they argue that the heuristics approximate the normative models quite well most of the time, despite their flawed logic, and it is only under unusual circumstances that they result in error. For example, frequency estimates based on availability, or "the ease with which examples or instances come to mind," should generally be accurate since frequency and memorability should be highly correlated. It is only in cases when memorability and frequency diverge, for example, when a risk is reported more often in the newspaper than its true frequency would dictate (see, e.g., Lichtenstein, Slovic, Fischhoff, Layman, & Combs, 1978), that relying on the heuristic results in error. However, due to the focus of researchers on locating and studying errors, empirical demonstrations of situations in which heuristics work well are absent. (The reader interested in historical parallels will find the same developments in the history of Gestalt psychology in relation to its focus on perceptual illusions; see Brunswik, 1956.)

Significance for the Study of Stress

The heuristics and biases research program provides an advantage over the decision analysis research program for studying the effects of stress on cognition because it provides a richer set of descriptive processes, specifically the three principal heuristics described above. For example, prospect theory might predict that, under stress, people may shift the reference point they use to assess gains and losses. Or contingent weighting theory might predict that, under stress, people may become more or less likely to weight inputs, depending on their compatibility with the required outputs. Yet the heuristics and biases approach is less helpful for evaluating the effects of stress on cognition, because it does not predict which heuristic will be used in which situation. It is possible, for example, that under stress people would pull an entirely different heuristic out of the cognitive toolbox than they would under normal circumstances, but there is no way of predicting which one.

With respect to the coherence of the process, the heuristics and biases approach would predict that, since coherence is so poor even under the best of circumstances, it can only get worse under stress: People employing simplifying rules of thumb are unlikely to switch to highly analytical rational or normative strategies under stress. Indeed, the conventional wisdom is that the reverse occurs. With respect to the approximation of the right answer, the heuristics and biases approach might predict either a decrease in accuracy or no change in accuracy, depending on the situation. For example, the stressor might change the situation from one in which a heuristic works well, yielding a close match between the judgment and the correct, calculated answer to one of those situations in which the heuristic breaks down. Or the person may already be in a situation in which the heuristic performs poorly, so the stressor may have little or no effect. It is difficult to imagine a situation in which a stressor fortuitously changes the circumstances from one in which a heuristic that has been performing poorly to one in which it performs better. It is also important because, as the reader will recognize, another way to put this would be to say: Negative events are more likely than positive or neutral ones to elicit "*coherent,*" that is, responsible cognition. Why coherent, responsible cognition? Because causal reasoning is able to answer the question "why?" The answer is: "because *this* leads to *that*." Admittedly, the answers in many cases will not go far, may not be able to present a wide-ranging, interlocking set of explanatory variables, but efforts to answer the "why" question will take us on a path that will appeal to coherence.

Taylor's (1991) review contains a number of provocative suggestions. For example, she suggests that negative events demand a coherent explanation, a suggestion that is strengthened by her conclusion that "there is some evidence that the search for a causal explanation for negative events is not merely a response to the need to predict and control that event and similar events in the future, but also to explain away the event in a manner that has few lasting implications" (p. 73). And that is important too, for "explain away" suggests that the explanation not only applies to the behavior of the events but also to the behavior of the person—presumably the operator. In other words—"I didn't do it! It wasn't my fault. The machine broke down!"—a response to negative events that will not be wholly unfamiliar to the reader.

Thus, Taylor's observation that "negative events are more likely than positive or neutral ones to elicit causal reasoning" is important because it indicates that negative events evoke thought; that is, negative events stimulate the search for a coherent explanation. Concomitantly, her observation also implies that positive events evoke complacency—perhaps a complacency that is undeserved. Surely, this is a strong hypothesis that deserves serious exploration.

Taylor's review also includes references to work on judgment and decision making in "emergency" situations. For example, "Thus, a strong rapid response to negative events, coupled with a strong and rapid diminution of the impact of those events, may be most effective for the organism in both the short term and the long term" (p. 79). This statement also deserves our attention because Taylor then suggests that "the initial response may enable the organism to overcome *positively biased* [italics added] thought processes to *deal effectively with the emergency* [italics added]" (p. 79). Thus, Taylor apparently is suggesting that negative events may "overcome" and thus displace "positively biased thought processes" which presumably are a danger to the organism. In short, negative events are not all bad; they may be bad for the task system, but good for alerting the operator.

Taylor goes on to say, however, "*muting* [italics added] . . . the impact of the negative event may be essential for the restoration of positive biases that appear to facilitate effective functioning in non-threatening environments" (p. 79). This suggests that "positive biases" are acceptable after all. Indeed, they "appear to facilitate effective functioning in non-threatening environments." Although this argument may be difficult to unravel, it is nonetheless important because Taylor is pointing to the possible impact of "negative," that is, stressful, events on cognitive activity in a manner not ordinarily considered.

Taylor also suggests that there is an "offsetting response to arousal, which occurs automatically as a compensatory process that reverses its [arousal] effects" (p. 72). Could that be interpreted to mean that operators at a control site notice a negative event that occurs briefly, but fail to respond because of a compensatory response ("it didn't happen")?

Taylor clearly is breaking new ground in important ways that carry significance for the study of the effects of stress on judgment and decision making.

Research at the University of Illinois' Aviation Research Laboratory has begun to focus on the effects of stress on various heuristics and biases (see, e.g., Wickens, 1987; Wickens & Flach, 1988; Wickens, Stokes, Barnett, & Hyman, 1988; see also Raby, Wickens, & Marsh, 1990; Stokes, Belger, & Zhang, 1990). Raby et al. (1990) investigate the role of various cognitive biases introduced from the heuristics and biases research program with regard to the effects of workload pressure. They consider such topics as planning/participation, overconfidence in time estimates, cognitive "leveling" and task prioritization, and task tunneling. All their results are worthy of attention from stress researchers.

A detailed description of the flight simulator used at the Aviation Research Laboratory is provided by Stokes (1991) in *Human Resource Management in Aviation.* In the same volume, Jorna and Visser discuss the effects of anxiety on performance in a flight simulator. They report that their

> study investigated the occurrence of anxiety, and its effects on the performance of high- and low-anxiety subjects. Instructor ratings were compared with objective measures of flight control. State anxiety was found to be increased, particularly by a differential response of subjects to removal of feedback. Anxiety did not influence deviations from the flight-path, i.e., maintaining heading and altitude, but aileron control proved to be quite different. A surprising result was that instructors significantly favoured the high state-anxious subjects. These higher ratings were not supported by objective measures of performance. (1991, p. 123)

Information Integration Theory

Information Integration Theory (IIT) was developed by Norman Anderson in the early 1960s (for an overview, see Anderson, 1979, 1981, 1982, 1996; for recent developments, see Massaro & Friedman, 1990). Anderson's work grew directly out of early work in psychophysics that sought to identify simple mathematical rules to describe the relationship between stimuli and perceived sensations. For example, the physicist Gustav Fechner (1860/1966) was the first to propose a lawful relationship between a measurement of a physical stimulus (say, the loudness of a sound) and a measurement of an inner mental experience (the perceived loudness of the sound). The relationship (known as Fechner's Law, now replaced by Stevens' Power Law) takes the form $S = k \log I$, where S is sensation, k is a constant, and I is the physical intensity of the stimulus, and it can explain why ever larger outputs in stimulus intensity are required to obtain corresponding changes in perception.

IIT attempts to do precisely what Fechner did in psychophysics, but for a broader range of more complex tasks in cognition and social perception, that is, to examine how people subjectively evaluate the stimuli that impinge upon them and to uncover the mathematical rule or "cognitive algebra" that describes how these perceptions are *integrated* into a response or judgment. For example, when studying impression formation, one can ask under what conditions subjective ratings on different attribute dimensions are added together or averaged together to form an overall impression of a person (Anderson, 1965). In order to do this, IIT places considerable emphasis on measuring, in a precise metric sense, the social as well as the physical judgments of people, and therefore also owes an intellectual debt to such pioneers in psychological measurement techniques as Thurstone (1931).

Experiments in the IIT tradition rely on the concept of "functional measurement," which presupposes that psychological values are measurable and can be represented numerically. A typical experiment focuses on measuring the psychological values of the stimuli and the psychological value of the response or integrated judgment. By having subjects evaluate several targets with stimulus values that are manipulated in a factorial design, the pattern of responses can be compared to the pattern that would be expected from various algebraic rules such as addition, averaging, or multiplying. Anderson and his colleagues have used this paradigm to demonstrate that simple algebraic models can predict judgments in a wide variety of complex tasks. For example, Anderson (1962) showed that impressions of likableness could be described as the sum of the subjective values for two adjectives that described a person. Shanteau (1974) had people judge the worth of lottery tickets with different probabilities and values, and was able to show that their judgments followed the multiplicative law of subjective expected utility: subjective expected utility = subjective value × subjective probability. And Lopes (1976), in a study of betting behavior in five-card stud, found that subjects' subjective probabilities of winning followed a conjunctive multiplication rule: The subjective probability of beating two opponents, A and B, was well described by a formula that simply included the subjective probability of beating A times the subjective probability of beating B.

There is a resemblance between IIT and SJT because in both theories experimenters present a number of profiles of cues to the subject. But the resemblance ends there. IIT is a coherence theory (whereas SJT is a correspondence theory) because IIT focuses on discovering the coherence of the organizing principle that the subject uses to integrate the information received from the data display. Thus, IIT researchers wish to discover whether the subject coherently integrates information by averaging, adding, or multiplying the data, and the extent to which such models (adding, averaging, multiplying) account for the subject's responses. In general, these researchers are not interested in the empirical accuracy of a subject's judgment of some object or event; it is the coherent integration of information that is the center of their attention. SJT researchers, on the other hand, give much more attention to judgmental accuracy.

Significance for the Study of Stress

Since IIT researchers are interested in measuring subjective values and judgments and testing algebraic relations between them, they provide a wealth of parameters and process components in a variety of domains that might be affected by stress. One advantage of the ITT approach is that it can provide baselines against which the effects of stress can be measured, even when there is no established normative model. For example, there is no accepted "normative" procedure for forming an impression of another person; yet, judgments can be measured against a well-specified additive (or other) model.

IIT suggests that stress can have two effects on a given judgment task. First, stress could change the subjective value or judgment attached to different pieces of incoming stimulus information. For example, consider the Lopes study of betting behavior: Stress could affect the subjective value of beating any particular opponent. Second, stress could affect the integration rule by which subjective stimuli are combined: Under stress, the subjective probability of winning in the Lopes study may no longer closely follow a multiplicative function of the individual probabilities, but might diverge from the multiplicative rule due to errors, or even follow an entirely different algebraic rule. Moreover, these two effects are independent: It is entirely possible, for example, that stress may change all of the subjective stimulus values but have no effect whatsoever on the algebraic relationship between them.

So far as I can ascertain, IIT has not been employed in the study of the effects of stress on judgments.

Payne, Bettman, and Johnson

John Payne has been a pioneer in the development of methodology and theory in the study of multiattribute choice for both risky and riskless options (Payne, 1976, 1980; Payne & Braunstein, 1978). In a typical multiattribute choice study, subjects are asked to choose between several alternatives (say, different apartments for rent) that vary on several attributes (rent, distance from work, etc.). Information about attributes is concealed behind the windows of an array, and subjects reveal the information they are interested in as they go about making their choice. By observing the order in which information is revealed and collecting think-aloud protocols, the cognitive process or processes underlying the choice can be determined. Subjects may, for example, search all the attributes for each item before going on to the next item, search all the items on a single attribute and then go on to a second attribute, or adopt any number of simplifying strategies such as elimination by aspects (Tversky, 1972), satisficing (Simon, 1955), equal weighting (which ignores probabilities for risky alternatives), or a lexicographic rule which focuses on comparing only the most important attributes.

The observation that individuals may employ a variety of different strategies for any given choice problem (Payne, 1982; see also Abelson & Levi, 1985) led Payne, Bettman, and Johnson (1988) to develop and test an adaptive theory of strategy selection for risky choice, in which problem complexity and time pressure (both of which may be considered to be stressors) play a large role in determining which strategy will be selected from the cognitive toolbox. In their model a decision maker attends to both the expected costs (primarily the mental effort required to implement the strategy) and expected benefits (primarily the ability of the strategy to make the best choice) of available strategies during strategy selection. In any given decision environment it is proposed that people will choose the strategy that maximizes coherence while minimizing effort. Therefore, if people have available multiple strategies which are more or less equally accurate, they will adopt the one that requires the least effort (i.e., the one that is most efficient). The model therefore, unlike heuristics and biases theories, points to, and provides a method of predicting, which heuristics are most likely to be employed in a given situation (although prediction of exactly which heuristic is most likely to be used requires knowledge about the decision maker's relative values for accuracy and effort). It is also proposed that the coherence of different strategies varies widely, depending on such task characteristics as complexity and time pressure, and that people are adept at recognizing these differences and at switching from one strategy to another as task conditions warrant.

To test their theory, Payne, Bettman, and Johnson (1988) ran a computer simulation that compared the relative coherence of 10 choice strategies (ranging from compensatory strategies which consider all the information to simplified noncompensatory strategies such as elimination by aspects and lexicographic strategies) as complexity (the number of alternatives and attributes) and time pressure were varied. The simulation was run on a variety of risky choice problems, in which subjects chose between gambles with multiple possible outcomes that varied in probability. The simulation demonstrated that, under time and complexity constraints, several heuristic strategies provided more coherence than a truncated normative strategy (which tries to integrate all the information but runs out of time). In addition, the coherence of any given heuristic was shown to depend on the structure of the decision problem (e.g., the dispersion of probabilities across attributes or the presence of dominated alternatives). No single heuristic worked well for the entire space of potential choice problems, demonstrating that multiple strategies are necessary to achieve coherence in many different situations and contexts. A companion study (Payne, J. et al., 1988) of human decision makers found that people made choices consistent with the patterns of efficient processing identified by the simulation.

Significance for the Study of Stress

One implication of the Payne, Bettman, and Johnson approach for the study of decision making under stress is that their work suggests a new baseline for examining accuracy. The normative criterion is no longer accuracy at any cost but "adaptiveness." Choice behavior under time and complexity constraints should not be compared to the full rational model but to a truncated one. Normative baselines are good criteria for accuracy only when time and mental resources are unlimited and when tasks are simple. People may therefore perform less well under stressful circumstances than under ideal ones but still be performing as well as they possibly can, given the environmental constraints.

The Payne, Bettman, and Johnson framework also provides an explanation for perceptual narrowing and filtration of information, which has often been suggested to be the most typical response by decision makers to stress (see Yates, 1990). People examine less information because it may be adaptive to do so. Those strategies that process all the choices on only a limited number of attributes (e.g., the elimination by aspects or lexicographic strategies) yield the most accurate judgments under time pressure. For choice problems, staying with the normative model (which requires processing all the attributes for all the choices) as time pressure increases would not be the optimal strategy. Thus, we find again empirical results that deny the proposition that "narrowing" leads to flawed decisions.

A close examination of the time pressure results of Payne, Bettman, and Johnson indicates that there may be a "hierarchy" of responses to stress. Under slight time pressure, their subjects stayed with the normative compensatory strategy but tried to process information faster. As time pressure increased to moderate levels, they began to narrow their focus, looking at only part of the available information. Finally, under severe time pressure, they completely abandoned the compensatory strategy and switched to heuristic shortcuts.

The Payne, Bettman, and Johnson approach is similar to the heuristics and biases approach in that it assumes that coherence is usually poor but that empirical accuracy is often quite good anyway. Under time pressure, however, coherence is lost. Overall, their model is one of a decision maker who is both coherent and accurate, given unlimited time and mental resources and simple problems, and who adapts quite well to stress by filtering information and changing strategies. Although coherence decreases relative to a full normative model, it is maintained relative to the best that can be done under the circumstances. However, it should be noted that they have thus far studied only a single judgment task, multiattribute choice, which people perform very well when not stressed. It is not clear how their framework would apply to tasks which people typically perform poorly (or even less than optimally) under the best of circumstances.

Part II

NEW DIRECTIONS

Part I contains the serious charge that although much research has been carried out on the effects of judgment under stress, it has not been organized into a useful corpus of knowledge—nor can it be. Despite the fact that many excellent studies have been carried out, the work is splintered over at least four different fields, and even within the field of judgment and decision making it is divided into two separate fields that are not in contact with one another. In addition, the concept of stress has never been sufficiently removed from idiosyncratic definitions to enable it to be of use across research programs. Indeed, its definitions have been arbitrary and divorced from theory. As a result of such diversity of definition, method, and result, neither academics nor those facing problems of preparing for judgments under stress are likely to be helped by the work that has been done. Change is needed.

Therefore, part II presents my reformulation of the problem of understanding judgment under stress. I begin by presenting seriatim theories of stress, judgment, judgment tasks, of how cognition and task interact, and how that interaction is affected by what is commonly known as "stressful" conditions.

6

A Theory of Stress

In this chapter I present a new concept of stress in the attempt to remove the present vagaries that now make it so difficult to use as a research tool.

Origins of Stress

I begin with two premises: (1) All organisms attempt to maintain *stable* relations with their environment, and (2) the *disruption* of stabilized relations produces phenomena known to lay persons and professionals alike as "stress." (A caveat: Under conditions called "play," certain organisms, notably human beings, seek circumstances that counter these premises.) The *first* premise can be refined, made specific and operational, and non-circular measures can be developed, by reference to the empirically well-established phenomenon of *constancy.* This term refers to the organism's ability, for example, to estimate correctly the size or shape or color of an object despite changing sensory inputs. Thus, the size of an object remains *constant* to us no matter how far away it is. Similarly with sound; even though a very loud sound may be heard only faintly because its origin is far away, we are still able to detect it as a loud noise, an explosion, for example.

The *second* premise—that disruption of stabilized relations produces "stress"—leads directly to testable propositions regarding a person's cognitive responses to a disruption of constancy. And that is all we need; the ambiguities in the lay (and professional) concepts of stress, and the numerous vague, and often circular, professional and scientific concepts of stress, can be set aside in favor of concentration on a specific, well-known form of behavior—maintenance of constancy. Analysis of its *disruption* and its *consequences* will be our topic. Our goal is to replace the confusing connotations of the word "stress" with reference to specific forms of behavior within the context of empirically well-established phenomena, namely, the phenomena of "constancy." I now turn to a further elaboration of why I chose this form of behavior as a point of departure for studying judgment under stress.

Constancy

Constancy is one of the fundamental discoveries of psychology, if not *the* fundamental discovery. Constancy has been found to exist over a wide range of perceptual processes, including shape, color, brightness, density, weight, volume, tactile-kinesthetic sensations, loudness, and even values (see any text on perception).

For most psychologists, constancy is merely an interesting phenomenon, a challenge to find out exactly what the mechanism of this remarkable ability is. But for the psychologist Egon Brunswik, constancy was "the essence of life." As he put it, "stabilization effects of the kind studied in the . . . constancies are of the very essence of life" (Brunswik, 1956, p. 23). More specifically,

> stabilized relationships with the environment are biologically useful adjustments, especially when they anchor organismic orientation to properties of more or less remote, more or less vitally relevant solid objects of potential manipulation and locomotion such as landmarks, tools, enemies, or prey which themselves are usually fairly stable or predictable. (p. 23)

What he meant by this is that when organisms such as mammals orient themselves to the "remote" properties of solid objects ("landmarks, tools, enemies") rather than their immediate properties that produce sensory stimuli (light, heat, pressure), the organism is more likely to be selected by its environment because it is more likely to be effective in it and thus to survive, to become more ecologically fit, and thus more likely to pass on its genes. This strong claim was made in the 1950s when scientific psychologists were still in the grip of a stimulus-response metatheory that sanctified exactly those sensory stimuli—to be studied one at a time—as the only appropriate materials for consideration; anything else was too "soft" to be appropriate. The cognitive revolution of the 1960s and 1970s changed that narrow stimulus-response conception of acceptable and proper psychological inquiry. Indeed, Brunswik's premise that the organisms' behavior was oriented toward "stabilized relationships with . . . vitally relevant solid objects of potential manipulation and locomotion," rather than proximal sensory stimuli could be said to characterize that revolution. And it is precisely that premise that I will use to support my pursuit of judgment under stress. For I intend to argue that constancy is so central to our existence that the ***disruption*** of constancy, that is, disruption of stabilized relations between an organism and those "more or less vitally relevant solid objects," presents a threat to the organism that induces not merely an affective response, but a cognitive one that is intended to *reestablish* [italics added] stability and thus survival (see Mandler, 1982; Simon, 1979, pp. 33–36, for related ideas). Moreover, I shall expand the concept of constancy beyond "stable relations" with physical objects to "stable relations" with social objects such as people.

My assertion of the (largely overlooked) importance of the constancy phenomena receives considerable support from the recent neurobiological literature on stress. For while the idea of an organism achieving a stable interaction with its environment despite its change—and its change on the sensory surface of the organism—is described by psychologists as "constancy," it has a clear counterpart in the neurobiological literature where it is described as "allostasis." This term was apparently introduced in 1988 by P. Sperling and J. Eyer in an article titled " 'Allostasis': A New Paradigm to Explain Arousal Pathology" and was intended by them to mean "stability through change" (p. 636). None of the 46 other authors in the 700-page *Handbook* in which their chapter appears uses this term, however, and it has yet to find its place in a dictionary or common usage. As far as I can ascertain, it was next used by B. McEwen and E. Stellar in 1993. It was then used by McEwen (1998) as the central concept in a review article in the *New England Journal of Medicine* to mean "the ability to achieve stability through change" (p. 171), a definition that can be applied directly to the constancy phenomena. The parallel with the term constancy becomes even more apparent when McEwen points out that this ability is "critical to survival," a feature also noted by Brunswik in relation to constancy when he wrote that "stabilization effects of the kind studied in the . . . constancies are the very essence of life" (1956, p. 23).

The critical significance of "allostasis" as a new idea was emphasized by Sperling and Eyer when they contrasted this concept with *homeostasis:*

> Allostasis involves the whole brain and body rather than simply local feedbacks, is a far more complex form of regulation than homeostasis. Yet it offers definite advantages. One is that it permits a fine matching of resources to needs. In homeostasis, negative feedback mechanisms, uninformed as to need, force a parameter to a specific "setpoint." If blood pressure were actually determined in this way, that is, set to an average, "normal value," it would almost invariably be too high or too low for whatever was going on at the moment. Allostasis provides for continuous re-evaluation of need and for continuous readjustment of all

> parameters toward new setpoints. This makes the most effective use of the organism's resources. (1988, p.637)

Thus, in neurobiology the allostasis view of organism-environment stabilization is exactly the same as that implied by the constancy phenomena in psychology. And homeostasis is rejected by—at least some—neurobiologists as an adequate explanation of the behavior of the organism for the same reason that the stimulus–response view of behavior has been rejected by those psychologists who have participated in the "cognitive revolution." Within that "revolution" stabilization of the organism with points remote from the organism is also achieved by ignoring "local feedbacks," that is, different sensory stimulation on, say, the retina, as the object changes its form or distance. In both cases, localized, specific stimulus points become irrelevant because of their loose connection with both the object and its perception.

Moreover, Sperling and Eyer place the same emphasis on "disruption" of stabilized relations as I do, and indeed, they present considerable empirical evidence indicating how such disruption works at both the neurobiological and the socioeconomic level. For example, they note that "age-specific death rates rise when intimate social relations are *disrupted*," that "*disruptions* . . . including war, migrations, and economic development, affect most strongly youth entering the labor market," and that "a large cohort entering the labor market during an economic expansion experiences greater competition and social *disruption* and has correspondingly higher mortality" (p. 629, italics added). They further disparage the ability of current theory to provide explanations for these events by pointing out that textbooks of physiology offer no explanations for them: "No text explains . . . why in modern society blood pressure rises with age. . . . Nor do they explain why this rise starts at the age when children enter the environment of school," and they offer a remarkable set of statistics in this regard. Further, "Texts do not explain why blood pressure is highest and hypertension most prevalent where social disruption is greatest" (p. 630). Their discussion of disruption at the neurobiological level focuses largely on the effects of arousal and is too technical to be included here, but the emphasis is the same. The orientation of the organism is directed toward maintaining stable relations with the environment, and it is disruption of those stable relations that should provide the independent definition of stressful circumstances.

Constancy and Evolution

Correspondence Constancy

There is a further reason for using the constancy phenomena as a point of departure for the study of the effects of stress on human judgment, and that is because of the close relation between the constancy phenomena and evolution. Natural selection, it can be supposed, would favor those creatures who possessed or developed correspondence constancy. Indeed, it is hard to think of a more critical behavioral feature, one more integral to the survival of those early hominids and Homo sapiens, than the ability to estimate correctly the size of another animal irrespective of its distance, or the ability to estimate the loudness of a sound irrespective of its distance, or the ability to see that a certain object is of a certain shape, round, say, or square, no matter how it is presented, no matter what shape is projected onto the retina. Correspondence constancy must be as central a part of our inherited biological makeup, and as valuable in establishing our ecological fitness, as binocular vision.

It is also reasonable to assume that correspondence constancy has been with us for as long as Homo sapiens have existed as a species. If this is true, then it is not only a remarkable ability, but also one that we should expect to be highly resilient to external trauma. That is, correspondence constancy should be highly resistant to disruption from such naturalistic events as heat, cold, high winds, blows to the body, and similar disruptive events. Loss of correspondence constancy should occur to human beings only under the most extreme conditions, perhaps only when consciousness itself is lost. In short, only rarely should we lose correspondence constancy under any natural circumstances. Indeed, the extraordinary steps that are required to deprive persons of correspondence constancy can be seen in the NASA procedures (e.g., the centrifuge) to test and train astronauts.

That conclusion has profound implications. It argues that although Captain Haynes and his crew were severely tested by the conditions that confronted them—the irregular and uncontrollable changes in the altitude of the aircraft—they never lost their orientation in three-dimensional space. Correspondence constancy was maintained. Nor did the extreme physical conditions of change, fire, explosions, noise, and heat, as faced by Commander Rinn and his crew on the *Samuel B. Roberts,* disrupt their correspondence constancy. Therefore, whatever stress these officers faced, it did not arise from a loss of correspondence competence for the reason it was *not* lost.

It is of considerable importance that both Captain Haynes's and Commander Rinn's reports of their reactions to their situations provide ample support for the conclusion that there was no decrement in correspondence constancy in their performances, nor, apparently, in the performance of their crews. Clearly, their correspondence constancy was highly resilient to the strong external assaults on it. Stress must have come from another source, for we know that both men, Commander Rinn on the bridge of the *Samuel B. Roberts* and Captain Haynes at the controls of UAL 232, did experience stress (in the lay sense of that term).

To be certain that the reader understands what is intended here, let me explain in different terms. Neither man lost his orientation in three-dimensional space, nor did either lose his ability to make accurate judgments of the size of objects despite variation in their distances, or lose his ability to detect the shape of objects, nor did they lose their ability to differentiate between loud noises heard at a distance and softer noises heard nearby; they maintained their correspondence competence. But each lost, in part, his coherence competence. Commander Rinn found that if he followed naval doctrine, his ship would be lost, and he had to find a new coherent plan of action, which his executive officer challenged as incoherent and thus foolhardy. Captain Haynes never regained a coherent control system for his aircraft because it had to be repaired, and that couldn't be done; therefore, he had to rely on his inherent ability to make use of multiple fallible indicators to provide him with the correspondence competence that allowed him to guide the airplane to the airfield. Before we accept that conclusion, however, we must consider in more detail the nature of *coherence* constancy, and to that we now turn.

Coherence Constancy

The term "coherence constancy" refers to a person's ability to apply the same *coherent principle* to a problem, despite its various forms of display or presentation. This definition is thus analogous to that for correspondence constancy; both refer to a person's ability to infer correctly an underlying, that is, not immediately apparent, condition, despite a changing proximal display. That is, our judgment of an underlying condition is *constant*, despite its changing appearance. Do we have the ability to achieve coherence constancy intuitively? We surely do. For example, coherence constancy can be seen when we recognize the same melody played in different keys; thus, all the coherence of the melody (the fundamental relation among the notes, what we recognize as a "melody") is preserved for us, despite the change in the specific notes (the change in sensory information) that convey the melody. In fact, the Gestaltists demonstrated again and again our drive and capacity for intuitive coherence through their presentations of perceptual illusions. It was through the use of illusions that they showed that people (often, if not usually) see, not what is actually physically present, but what *ought* to be present in a coherent form of an object (a circle that is physically incomplete is *perceived* to be complete). There is little dispute about our intuitive coherence competence. The Gestaltists demonstrated that quite clearly; that is why their illusions of coherence are in all the textbooks.

But competence in *analytical* coherence is another matter; it must be *taught*. We do not have an intuitive grasp of the fact that adding a constant to an equation preserves the relation among the variables, despite producing different numbers. We also have to be taught that a syllogism's logic should remain the same, despite changes in the symbols used to present it. Some changes are easier than others, of course. For example it is easy to see that our premise remains the same whether we say "if all x's are y's," or we say "if all a's are b's." But if we change the nature of the information significantly, change the letters to things, particularly things we know about, then it is not immediately apparent that the premise remains the same. And, as a result, those untutored in logic make mistakes. (See Hammond, 1996, for such elementary failures in coherence constancy in Supreme Court Opinions.)

Can the concept of "coherence constancy" be used in the same manner as the concept of "correspondence constancy"? Can it be said that "stabilized [coherent] relationships with the environment are biologically useful adjustments," as claimed for correspondence constancy? Yes; in both the intuitive and analytical sense. The Gestalt psychologists emphasized the intuitive component of coherence, so much so that they claimed that coherence constancy should have the same, if not greater, claim to our attention as correspondence constancy. Their reason for that claim was that they believed that the world, both physical and social, is organized in terms of coherence, or, as they put it, "wholes" or "patterns," that is, "gestalts," and that our brains and cognitive processes are organized in parallel fashion. And, indeed, it is that parallel organization that accounts for our intuitive ability to cope with the world. In short, in the Gestaltists' view coherence reigned, in both the outside world and our intuitive cognitive activity.

This view of the dominance of intuitive coherence has largely disappeared from modern psychology, aside from its legacy of perceptual illusions that are a staple of every introductory textbook. Yet we should ask: Aren't the demands for coherence constancy greater now than in the time of the hunter-gatherer who depended almost entirely on correspondence constancy? I think the reasonable answer is—yes. Coherence constancy is an important element of a stable existence in those regions of society—the worlds of business, engineering, science, indeed in the everyday life of anyone who sits in front of their personal and/or work computer. It is becoming indispensable. Many otherwise incoherent environments have been engineered into coherent forms that demand coherent cognition to a degree never before imagined. It is surely true that in modern societies, coherence constancy, that is, getting the logic of the problem correct despite its various forms of symbolic presentation, is indeed now survival related. Thus, the intersubstitutability of symbols in coherence constancy becomes parallel to the intersubstitutability of *stimuli* in correspondence constancy.

Coherence competence is certainly necessary on the flight deck of an airliner, or a ship on the high seas, or in any carefully designed control center, as for example, in the control of the launch of nuclear weapons, or a power-generating station, and the business and financial centers. In less esoteric circumstances, one must be able to see that distance remains the same whether measured in kilometers or miles, or that the schematic representation of subway routes or bus routes are simply symbolic representations of the geographical routes. Our current task environments are very frequently constructed, that is, engineered, in a manner that demands coherence constancy. The computer program, of course, offers the prime example of an environment that is absolute in its demands for coherence. Thus, coherence constancy may be as valuable in today's world as correspondence constancy was in the days of the hunter-gatherer.

It is curious that although coherence constancy has played an important and significant role in research on judgment and decision making, its role has not been identified as such; we do not find articles that refer to loss or maintenance of coherence. Yet one of the principal goals of the research program initiated by Bayesian researchers (see, especially, Kahneman, Slovic, & Tversky, 1982; Tversky & Kahneman, 1983, et seq.) has been to demonstrate the (almost) ubiquitous failure of the "principle of invariance," a principle that demands that judgments should remain the same, despite variation in the form of the presentation of the problem. (This is exactly the same process that is measured in the Raven Progressive Matrices test in visual form.) To take a simple example, the answer to "what is the sum of 2 + 2" should be the same as the answer to "what is the sum of 1 + 1 + 1 + 1." To go beyond simple arithmetic, the answer to the question "Is the probability of X given Y, the same as the probability of Y given X?" should be no. Thus, the principle of invariance is another way of describing coherence constancy.

Throughout some 20 years of research, researchers have claimed to find example after example of what is often called a "violation" of the principle of invariance, and therefore a violation of coherence constancy. (See Plous, 1993 for examples.)

An often-used example from Tversky and Kahneman (1981) is the following:

One group of subjects is given this problem: Imagine that the United States is preparing for an outbreak of an unusual Asian disease, which is expected to kill 600 people. Two alternative programs to combat the disease have been proposed. Assume that the exact scientific estimate of the consequences of the programs are as follows:

> If Program A is adopted, 200 people will be saved.
>
> If Program B is adopted, there is ⅓ probability that 600 people will be saved, and ⅔ probability that no people will be saved.
>
> Which of the two programs would you favor?"

A second group is given this version:

> If Program C is adopted 400 people will die.
>
> If Program D is adopted, there is ⅓ probability that nobody will die, and ⅔ probability that 600 people will die.
>
> Which of the two programs would you favor?

It will be apparent that Programs A and C are equivalent, as are Programs B and D; expected survival is the same in these cases. But a majority of subjects chose Program A over Program B in the first group and Program D over C in the second; thus demonstrating that changing the text resulted in a change of preference. Thus, the results were not invariant over formally identical conditions; coherence constancy was absent. Numerous studies of this sort demonstrated the fragility of coherence constancy in the context of probabilistic judgments—judgments under irreducible uncertainty. (The stability of these results has been seriously challenged, however. See Cosmides & Tooby, 1996; Koehler, 1996; Lopes, 1991; Lopes & Oden, 1991.) Ironically, then the invariance, or constancy, of these results have themselves been challenged.

In short, there are two types of constancy: one, correspondence constancy, that is well established, highly general over different psychological functions (vision, audition, etc.) as well as highly general over wide variations in the ecology (an ability shared with the apes and many other species); and a second type—coherence constancy—that is far less established in terms of its generality, and far less understood in terms of its biological and cultural determinants, despite its increasing importance.

Correspondence constancy has long been recognized by psychologists and indeed now appears to be taken for granted. The idea of correspondence constancy appeared in judgment and decision research in the beginning of this field (see Hammond, 1955) and has been pursued steadily. There are now hundreds if not thousands of studies of this topic that demonstrate the characteristics of correspondence constancy, partly because of the obvious importance of the empirical accuracy of judgments, and because the relation between correspondence constancy and survival and fitness is easy to see. Coherence constancy has also been studied on hundreds of occasions and in a variety of ways. It was set in motion by Edwards (1954) but the work by Tversky and Kahneman (1974, et seq.), that demonstrated the apparent *lack* of coherence considerably accentuated its visibility, and attracted attention not only from psychologists, but from philosophers and economists as well. To the latter groups this meant an attack on the *rationality* of Homo sapiens, a topic that has been under scrutiny from antiquity; the experiments by these psychologists added a new dimension and vitality to the discussion. Unhappily, the complementarity of the two metatheories of judgment and decision making has not been obvious to its students, and the connection between the two has not been made plain (but see Hammond, 1996).

They must be made plain, however, if we are to make progress with the investigation of judgment under stress. In what follows, I will make use of both concepts, and will demonstrate that it is the *disruption* of both types of constancy that should take center stage in the study of stress and human judgment.

Disruption of Constancy

The concept of disruption of constancy will offer freedom from the wide variety of detached, atheoretical suppositions regarding stress and judgment, as well as the numerous circular and nonfalsifiable definitions of stress that remain unique to specific investigators. And because the approach taken here embeds the situations and behaviors of interest in constancy—a well-known, well-documented, domain-independent aspect of human behavior in general, and human judgment in particular—the analysis of the causal sequence of events and the cognitive responses to them become straightforward. Investigators are thus afforded the opportunity to develop a cumulative discipline rather than one splintered over various "stressors"; they can build on a general phenomenon and empirical facts related to testable theories. Hypotheses regarding "disruption of constancy" can readily be reduced to testable, falsifiable predictions regarding observable cognitive responses to manipulatable, environmental, independent variables. In short, disruption of constancy offers a coherent researchable domain. And that, in turn, will make it possible to understand and predict judgment and decision making processes that are evoked when people seek to maintain constancy—a stable relation with their environment, physical or social—under circumstances that prevent, hinder, disrupt, or threaten to disrupt, their efforts to do so. Although constancy phenomena have been widely studied, the *disruption* of constancy has rarely been investigated (aside from sensory deprivation experiments which do not cast much light on the cognitive processes affected).

A first step in this direction is to differentiate the source of these disruptions.

Endogenous and Exogenous Disruptions

Although this distinction has long been made in the human factors literature it has not been employed in studies of judgment and decision making under stress. Vicente (1992) makes use of the idea of disruption when he points out that the appearance of unanticipated events for which there is no place in the design of the system "is the greatest threat to safety" (p. 543). In what follows I will refer to any disruption *within* the task system itself (e.g., loss of an information-providing instrument) as an *endogenous* disruption. Disruptions due to factors *outside* the task system—fire, noise, heat, cold, vibration—are here considered to be *exogenous* disruptions. Xiao and his colleagues (Xiao, Hunter, Mackenzie, Jefferies, & Horst, 1996), for example, make use of this distinction when they show that medical doctors resuscitating trauma victims in the emergency room were more stressed by task-related (i.e., endogenous) stressors such as workload, time constraints, and diagnostic uncertainty than nontask-related (exogenous) stressors such as noise and team interactions. As a test of the utility of this distinction I offer the following hypothesis: Overcoming *endogenous* disruption of constancy demands *cognitive change*, whereas overcoming *exogenous* disruption of constancy demands *resistance to* cognitive change. I illustrate these effects by two examples: The first illustrates *endogenous* task change that disrupts coherence constancy and induces cognitive change from analysis to intuition; the second illustrates the opposite, endogenous change disrupts coherence constancy and induces a change from intuition to analysis.

ENDOGENOUS DISRUPTION OF COHERENCE CONSTANCY

Endogenous disruption can result in a demand for change, even creativity, or a demand for suppression of change, or creativity.

Demand for change to intuition. Here I refer to innovation that results from a largely unconscious, nonretraceable cognitive process. Intuitive innovation implies that a new idea or activity has sprung rapidly from unknown cognitive activity, usually under time pressure. In what follows I use once more the example of Captain Haynes's predicament on UAL 232, for it provides an example of intuitive innovation—creativity—initiated by endogenous disruption of coherence constancy.

After the loss of control became obvious, the first step taken by the captain was to ask the experts on the DC-10 at the United base in San Francisco for advice. Within a few minutes everyone realized that the United base could not provide any information that would be of help; the situation of UAL 232 was not only unheard of, but also unbelievable. Indeed, the crew had difficulty in making the experts in San Francisco understand that *all* the controls were gone, without doubt an unanticipated event; the entire control system lost its organization; it became incoherent.

The situation was now indeed perilous; the crew was barely able to control the airplane and contact with the United base had proved to be useless. And because such an endogenous disruption had never been encountered before in the history of civil aviation, it demanded that a new method of flying the airplane be found immediately. The specific task facing the crew was to keep the airplane from flipping over onto its back. In short, *creativity was demanded immediately after an attempt at an analytical solution* by way of consultation with the experts at the United base failed.

Fortunately, the crew did discover a new method of flying the airplane; the crew *discovered* that appropriate orientation could be achieved and maintained by the innovative means of throttle manipulation. Moreover, they learned how to do it; they learned the (intuition-based) perceptual-motor task of flying the airplane by means of changing the application of power to the right and left engines. Part of the reason for learning was that they received immediate feedback of the most useful kind; they could directly perceive the consequences of their actions; increase power to this engine and the wing goes up; decrease power and the wing goes down—more or less. But it wasn't all that easy because of other characteristics of intuition.

One of the long-accepted primary characteristics of intuitive cognition is a lack of awareness of the use of information that is in fact controlling one's behavior. Another is a lack of awareness of the nature of perceptual-motor coordination and action. Both characteristics were present in the cognitive activity of the United crew. The captain of UAL 232 comments on both in his lectures. He notes that although he knew intellectually that it was useless to try to fly the airplane by the customary means of manipulating the yoke, he could not stop moving it, even though he knew the airplane was being directed by the differential speeds of the left and right engines. The throttles were being manipulated by another crew member; but he could not remove his habitual dependence on the yoke; he continued to grip it tightly. As a further lack of awareness of the processing of information, Captain Haynes told the tower at the emergency landing field that the airplane could only make right turns. Yet the trace of UAL 232's flight path clearly shows a left turn on the final approach (as he explains in his lectures, a fact that still puzzles him).

In short, an extraordinary threat to stabilized relations, due to a unique, never-before experienced, wholly unanticipated endogenous failure that rendered the airplane's operating system incoherent, demanded that the crew create a new way of flying the airplane *immediately.* The crew did that; application of the properties of intuitive cognition (e.g., unconscious use of multiple cues) shows that it was creative intuitive cognition—together with extraordinary calmness and courage—that saved the airplane and most of the passengers on board. There is no need to appeal to the ambiguities of "stress" in this situation; all we need is to specify the loss of coherence in management due to an endogenous disruption of the control system and to note its cognitive consequences, specifically, the successful change from normal analytical cognition to intuition.

The UAL 232 episode thus strongly suggests that it will be useful to study judgment under disruption—or threatened disruption—of stabilized relations with the environment due to endogenous factors, with careful consideration being given to the process of *creativity on demand.* In the case of endogenous change under benign circumstances—an electronic or mechanical failure within the task system—the person's environmental supports (colleagues, handbooks) and personal resources (knowledge, experience, training) will ordinarily provide the necessary cognitive means for regaining coherence rapidly. If, however, neither availability of analytical assistance (manuals, expert advice) nor personal resources (knowledge, memory, previous training) enable one to combat destabilization, as in the case of UAL 232, then innovation and learning, as exhibited by the crew of UAL 232, will be demanded. But not all disruptions demand *intuitive* creativity.

Inducement of analytical creativity. Norman Maclean's (1992) compelling description of tragedy in his book *Young Men and Fire* provides another extraordinary example of creativity on demand in response to endogenous disruption of coherence constancy. In 1949, in the early days of "smoke jumping"—the first parachute jump was made in 1940—

15 young men parachuted into the head of Mann Gulch in the Montana wilderness to put out a small fire at the bottom of the gulch where it meets the Missouri River. After safe landings, the crew proceeded down the gulch according to plan. Suddenly, however, the plan became irrelevant; for an endogenous disruption occurred. Instead of the minor fire the crew expected, the fire exploded. Thus, the task of "putting out the fire"—it was a different fire—changed drastically. The crew now found itself faced with a roaring inferno that began racing up the gulch toward the surprised smoke jumpers. Their only chance of escape was to try to *run* up the steep slope of the gulch to the top and go over its side. But it quickly became clear that their chances of success—and of living—were vanishing; the fire was rapidly overtaking them. When the foreman saw this, he did something that, as far as anyone knew then, or knows now, had never been done before. He lit a grass fire in *front* of him, ran into the center of it, and laid down in its ashes. It worked; the main fire went around him and he survived. Only two escaped by running; thirteen of the crew were caught by the flames and died on the slope or soon after, the worst disaster (until 1994) in the history of the smoke jumpers.

There is no doubt that the foreman *invented* his solution to his awful predicament on the spot. He substituted a new, coherent plan for coping with the fire. But the novelty of his solution meant that it had no credibility. His idea was indeed counterintuitive, but that made it incredible, for it was precisely intuition that was controlling everyone else's activity. Although the crew members were close enough to the foreman to see what he was doing (indeed, he called to them and urged them to get into his refuge), but after one brief look, they all refused. The foreman wanted the crew to *stay* in the path of the fire, a wholly counterintuitive plea, when intuition was telling everyone to *run!* As a result, the counterintuitive nature of his suggestion cost them their lives. The foreman explained later that he heard one crew member shout, "To hell with that, I'm getting out of here" (p. 95), but that man perished. As a result of his analytical creativity, however, the foreman survived by *thinking out* this innovative solution that required him to act directly counter to what intuition was demanding.

Summary. Reviewing these events at a safe distance, as we are, we can see that it is creative *change* in cognitive activity—from analysis to intuition—that reduced the loss of life on UAL 232, and from intuition to analysis that saved the foreman of the smoke jumper crew, and could have saved the entire crew.

In the case of UAL 232 the crew changed from the analytical pursuit of a solution, from slowly checking various parts of the operating system to rapid, intuition-based, perceptual-motor learning (which the captain had difficulty describing and recalling later). The smoke jumpers' foreman shifted from the intuitive pursuit of height on the slope to the analytically derived plan of building his own fire in the ashes of which he could—and did—escape from the big fire. The failure to shift to a counterintuitive solution cost all but two of the smoke jumpers their lives. In short, when *endogenous factors* result in a loss of constancy, a stable relation with one's environment, *change,* of either intuitive or analytical cognition, is necessary to reestablish constancy.

While it may be commonplace to note that when things have gone awry, change is a good tactic: this is not always true, as the next section will emphasize. And that in turn will show us why it is important to recognize the distinction between endogenous and exogenous disruptions.

Exogenous Disruption of Correspondence Constancy; Suppression of Creativity

The intrusion of noise, heat, vibration, overcrowding, obstruction of vision, and similar unwanted conditions are examples of exogenous disruption of constancy.[1]

In contrast with endogenous disruption, under exogenous disruption, with the operating or control system intact (endogenous disruption absent), an operator's task will be to *maintain* the application of his/her knowledge and skill, in short, to act as if the distracting conditions were absent.

In sharp contrast with the *endogenous* disruption that occurred to UAL 232, the Aloha airliner that lost a significant part of its fuselage during flight provides a good example of extraordinary *exogenous* disruption. Therefore, it provides a sharp contrast in the cognitive activity demanded of the crew. That is because the Aloha crew was required to *suppress* creativity, whereas creativity was *demanded* of the crew of UAL 232. Because the operating system of the Aloha aircraft remained intact, maintenance of correspondence constancy demanded that the crew employ their usual visual procedures for landing—dependence on visual constancy to align the aircraft with the runway, judge rate of descent, distance from the runway—despite terrifying exogenous factors (decompression and decomposition of the interior of the aircraft, as well as noise so loud that the captain and copilot had to use hand signals to communicate with one another). The successful suppression of creativity—*suppression* of a shift in mode of cognition from analysis to intuition—prevented disaster. Similar suppression of creativity also appeared in the behavior of the crew

1. So far as I can determine, Shanteau and Dino (1993) have carried out the only study that directly assesses the effect of stress on creativity. Their results were clear; stress, in this case, extreme heat, resulted in a definite decrement in scores on a test designed to measure creativity. Moreover, this measure of cognitive activity was the *only* one in the battery of 10 tests that showed any effects of heat-induced stress; it is noteworthy that none of the decision making tasks were affected.

of the United airliner that lost a cargo door shortly after leaving Honolulu.

The same resistance to creativity can be seen in Commander Rinn's testimony in relation to maintaining coherence:

> We deviated from standard doctrine, but the important thing was that the members of my crew, when I told them to do that, and when I directed through the chain of command, carried out the orders and carried them out emphatically and executed the training that they had received, they didn't stammer, they didn't make mistakes, they didn't run the P250s the wrong way, but they did exactly what they had to do on the basis of the training they had gone through and because things had been made very clear to them how they had to function. (Committee on Armed Services, 1989, p. 250)

Because Commander Rinn reports that he deviated from "standard doctrine," it appears that he is shifting cognitive activity. But, in fact, it is apparent from his testimony that he is shifting from one analytically derived plan (the doctrine that he should first put out the fire) to another analytically derived plan (dewater the ship first).

Note the critical difference between the foreman of the smoke jumper crew and Commander Rinn. The foreman thought of his plan on the spot; neither he nor anyone else had thought of it before and therefore his crew had received no training with regard to his counterintuitive creative suggestion; as a result they refused it. But Commander Rinn's crew "did exactly what they had to do on the basis of their training." Without such training we can surmise that they would have responded to Commander Rinn's orders as the smoke jumpers did to their foreman; they would have run from the scene to save their lives, as their intuitive judgment would have instructed them to do.

In short, if *exogenous* factors cause disruption, then established knowledge and training in the execution of established knowledge will enable a group to defy the disrupting agents ("stressors") and to continue to work toward maintaining constancy (Aloha) or restoring it (Commander Rinn). Training of both commanders and crew makes a difference with respect to the type of cognitive activity that will result from different forms of disruption of constancy and the success of leadership. (See Lee & Bussolari, 1989, for a report on the response of pilots with limited experience to platform motion in a simulator.)

Generality of the Concept of Disruption of Constancy as Stress

The introduction of the concept of disruption of constancy for defining the conditions of what is now termed "stress" is certain to raise questions about the generality of this idea. Constancy has heretofore only been applied to physical circumstances involving correspondence constancy. Extending constancy to include coherence constancy can readily be defended on the basis of theoretical argument and research results; it has already been done by means of the concept of "invariance"—certainly a legitimate application. But extending the idea of either form of constancy to social phenomena is entirely new. How confident should we be about the idea of "social" correspondence—or coherence—competence?

It seems to be plausible to assume that stability of social relations must entail correspondence competence—making correct judgments about events in a social group so that one's social behavior is guided accordingly—appears at first glance to be fundamental to social survival. The considerable research on the topic of the competence of social perception has produced conclusions that have varied from the position that we are hopelessly incompetent to being highly competent. The most convincing recent research is that done by Funder (1987, 1995, 1996), Funder and Sneed (1993), and Bernieri, Gillis, Davis, and Grahe (1996). It indicates that people can make accurate judgments about one another, but that accuracy is limited by the irregularity of the behavior of the persons being judged, rather than the incompetence of the persons doing the judging. As a result, empirical validity is fairly low. That is an important conclusion for it fits mathematically and empirically with other research on correspondence competence in the judgment and decision making literature (see chapter 3). In short, extending the idea of correspondence competency to social phenomena seems to be legitimate, theoretically and empirically, and to provide an integrative aspect to two very old fields of research. That conclusion leads us to inquire into the nature of judgment, social and otherwise. The theory presented below claims to be sufficiently general to include both coherence and correspondence types of judgments.

Applying the idea of coherence competence to social behavior also seems straightforward. It simply demands rationality of the sort we are all familiar with.

Summary

Disruption of correspondence constancy can require entirely different cognitive activity depending on task circumstances. *Endogenous* change that disrupts correspondence constancy will, if assistance is not available, require *change* in cognitive activity either from analysis toward intuition, or the reverse, depending on which constitutes the initial—now failing—mode of cognition. Disruption of constancy due to *exogenous* factors, on the other hand, requires *suppression* of urges to change the mode of cognition, despite intuitive demands. (Note the Darwinian theory at work here; the *environment* selects the appropriate cognitive activity for survival.)

7

The Cognitive Continuum Theory of Judgment

The theory of judgment to be described in this chapter is derived from Social Judgment Theory, outlined above in chapter 4. It is expanded here to form the Cognitive Continuum Theory (CCT) of judgment (see also Hammond, 1996). CCT is based upon four principal concepts; the correspondence and coherence metatheories (described in chapter 3) that control research in this field, and two theoretical terms—intuition and analysis (also described in chapter 3)—both of which have a long history. In the present chapter I first show the importance of the two metatheories in defining the different demands of judgment tasks, and then show the place of intuition and analysis within the two metatheories. I then treat the concepts of intuition and analysis in more detail.

The Importance of Distinguishing Tasks

Psychologists traditionally focus on the person's behavior and develop theories to account for it, and even to predict it. And that has led them to focus on cognitive processes and to ignore the problem of describing and differentiating among judgment tasks. Indeed, even the large differences between the judgment tasks within the two metatheories have not heretofore been made explicit. But these differences are critical. And that is because judgment is a *joint function* of task properties and cognitive properties. Therefore, we must have a way of describing tasks and differences among them, exactly as we must have a way of describing cognitive properties and differences among *them*. Our ability to evaluate the effects of disruption depends on our ability to do both.

In short, a theory of task properties is as necessary as a theory of cognitive properties; any approach that ignores either will be severely limited. In what follows I indicate the different demands made upon cognitive processes by different judgment tasks within each metatheory.

Demands by the Correspondence Metatheory

Social Judgment Theory (SJT) is one of the theories of judgment within the correspondence metatheory described in chapter 3. SJT is unique as a theory in that it directly addresses the problem of creating a theory of task properties intended to be general over correspondence demanding tasks. The theory is presented pictorially in the lens model (Figure 4.1) and presented mathematically as part of the lens model equation (see chapter 4). (It is not intended to be general over coherence demanding tasks, as, for example, in logic and mathematics.)

SJT requires that tasks be defined (at a minimum) in terms of (1) the number of cues (fallible indicators) presented to the subjects, (2) the strength of the relation between the cues and the criterion (the critical variable of the object to be judged), which is necessary for the measurement of empirical accuracy, (3) the degree of intercorrelation between the cues, (4) the form of the functional relation (linear, nonlinear) between the cues and the criterion, and (5) the degree of uncertainty in the predictability of the criterion, given the cue information. Notice that all these parameters refer only to the properties of the task. (Observation of the lens model diagram will make it apparent that it is symmetrical; task properties parallel the properties of the cognitive system. This symmetry is called the Principle of Parallel Concepts.)

Thus, correspondence tasks demand that subjects acknowledge the existence of the criterion variable to be judged, and attempt to judge its present, past, or future state. Achievement of empirical accuracy in our natural habitat, the ecology for which evolution has prepared us, requires nothing else from the subject. Thus, if you are requested to judge the distance of an object, you simply do it, you do not think about *how* to do it. Such tasks do not demand knowledge, thought, memory, training, awareness of the process, or any other cognitive capacity or ability other than that which is part of our biological heritage. In order to achieve empirical accuracy within any given substantive field of knowledge (e.g., meteorology, medicine), however, the task may require that the subject possess certain kinds of knowledge (which cues are most important), experience (accuracy in reading the cue information), and perhaps thought, depending on whether the task is inducing intuition or analysis. Accuracy may also require that the subject employ a specific organizing principle (make use of various patterns of data).

The important point is that this theory of tasks *requires* that the researcher describe the subjects' task in no more than these terms, but no less. And it is these terms that are sufficient to allow the researcher to differentiate among tasks. But without such a theory, the researcher is free to use whatever terms suit his or her whim.

One reason why the SJT concept of task properties is so important is that it tells us something about the subject's behavior, thus confirming the premise that judgment is a *joint function* of person properties and task properties. Thus, it tells us that in any correspondence demanding task the *limit* of empirical accuracy—the limit of correspondence competence—is controlled by the information in the task. For example, a forecaster's competence cannot exceed that allowed by the properties of the task. The best forecaster in the world—in any profession, meteorologist, stock broker, economist—cannot make accurate forecasts if the properties of the task will not allow it. And we need to know the limit of accuracy in a given task under *benign* conditions before we can ascertain the effects of disruption on accuracy. In the general case of forecasting we have some dependable knowledge regarding which task properties limit accuracy. For example, we already know, on the basis of both empirical and mathematical research, that the degree of irreducible uncertainty in the task is a limiting factor in the achievement of empirical accuracy (see Brehmer & Joyce, 1988 for examples; see Cooksey, 1996, for a recent and comprehensive review.)

But we need to know more about other task properties in relation to person properties; which task properties will affect the performance of which cognitive systems? Theory and research will guide the way to the answer to that question. For example, Chasseigne, Mullet, and Stewart (1997) examined the ability of elderly persons to learn to make use of *inverse* linear relations (instead of simple positive linear relations) between a cue and criterion. They found that the elderly had great difficulty in learning inverse relations from outcome feedback alone in comparison with learning positive cue-criterion linear relations, and that they were slower than middle aged persons. Thus, both person properties and task properties were examined in the same experiment because the researchers accepted the proposition that judgment is a joint function of person properties and task properties. Although disruption of stabilized relations was not examined in this study, it is easy to see that it provides an excellent point of departure for doing so.

In short, in naturalistic circumstances correspondence-demanding tasks require little competence from human (or other) subjects other than what evolution has provided; our judgments in these circumstances are empirically accurate, within the limits of the task situation as described by the theory. In circumstances that require more than what evolution has provided (task demands for coherence), learning will be required and that learning must be assisted either through experience or tutelage.

Demands by the Coherence Metatheory

The correspondence metatheory addresses empirical accuracy of judgments with respect to objective criteria. Things are different in the case of judgment tasks, in which the person is asked to make a judgment of the probability of an event that has never been observed and never will be observed. For example, what is the probability that the Japanese would have surrendered if the atomic bomb had not been dropped? There is no empirical criterion available to evaluate the competence of such a probability judgment as there would be if, within the correspondence metatheory, we asked a person to offer a probability of rain in the next 24 hours. Are there perfectly impeccable, perfectly coherent, theories (models, organizing principles, mathematical formulas) available by which to evaluate the process by which a person arrived at such a judgment? Yes; one frequently applied mathematical formula is Bayes' theorem (see chapter 5).

The fact that Bayes' theorem is a mathematical, statistical formula is highly significant, for it means that tasks demanding coherence do so by requiring a fairly high degree of technical competence for their solution, if their solution is to be attempted by analytical means. That is, if one is to provide a retraceable, highly justified process for reaching a probability that the Japanese would have surrendered (or any other similar event), then the task demands technical competence, and technical competence must be learned, somehow. It is also possible to provide a judgment for such matters by an intuitive, or quasi-rational means (and this is by far the most frequent case), but these judgments, insofar as they are not fully explicit, will not be fully justified, and as a result, will be disputed again and again. The fully justified mathematical solution may be disputed *because* it is a mathematical solution (analytical processes are still suspect in many quarters), but the process itself, if disputed, will be defended on analytical grounds. As a result, this dispute will be far different than the one generated by the intuitive judgment; in the latter there will be fewer appeals to logic and more to unappealing motives. In the former it will be the reverse.

Therefore, the first and most important distinction between the demands made by tasks within the correspondence metatheory and tasks within the coherence metatheory is that the former generally requires little or no justification and only in the case of the professions do they require training, whereas the latter require considerable justification, and thus training of the person, if they are to rise above dispute.

The importance of the above distinction between coherence and correspondence is that it enables us to see the difference in the demands made by the situation—the task—facing a person striving for correspondence competence and a person striving for coherence competence. The person striving for correspondence competence is faced with a task that demands empirical accuracy derived from the multiple fallible indicators it offers as information. The person striving for coherence competence is faced with a task that demands logical consistency derived from the abstractions that it offers as information. Thus, we can anticipate that a disruption of correspondence competence or coherence competence may have different consequences, a topic addressed in detail in chapters 8 and 9.

Intuition and Analysis Within the Two Metatheories

I began this chapter by reminding the reader that the approach taken here is based on four major concepts—correspondence and coherence, and intuition and analysis. In this section I take up the question of the relation among these four concepts. For example, one might ask: Can one achieve coherence *intuitively?* Yes. Can one achieve coherence *analytically?* Yes. And the same is true for correspondence: intuition and analysis are cognitive functions that can appear within this metatheory as well. I consider each separately.

Intuition and Analysis Within the Coherence Metatheory

Intuition

Our intuitive grasp of visual coherence—pattern—is easy to demonstrate; virtually everyone in the world knows what a circle ought to look like, and quite possibly, always has. Any five-year-old child can straighten a misshapen circle. The Gestalt psychologists labored mightily to teach us the importance of "good figure" in determining our perceptions; certain figures, usually symmetrical ones, just look good, and there is no need, or any possibility, of explaining to anyone why they do. We are all alike—although not unique—in this respect. But under special conditions our intuitive visual judgments can be wrong and misleading; something else the Gestalt psychologists insisted on teaching us. There are such things as *perceptual illusions,* as every student in the first course in psychology rapidly learns, and as the textbook writers never tire of demonstrating. Perceptual illusions can be created under special conditions, and they occasionally, but not often, occur in nature (the moon really does appear larger nearer the horizon). Most important, all of these intuitive illusions appear instantaneously, and they are compelling; in general, they are very difficult to overcome.

One also intuitively seeks coherence when telling a story, and the audience intuitively demands it. People tell stories to one another that are generally coherent without announcing or even being aware that they are seeking coherence. And, of course, the listener occasionally detects incoherence—a contradiction, an omission—in the story being told, and that is an embarrassment to the teller of the story. Seeking *incoherence* is frequently the role of the juror in the courtroom. Indeed, that is often the only test of truthfulness that can be applied. Thus, coherence that can neither be explained nor denied, and that is reliable across observers and over time can be found in intuitive perceptual judgments and conceptual judgments as well.

Persuasion often takes the form of presenting a "pattern" of evidence—in the form of numbers, actions, ideas—to people in the hope that they will be induced to make a certain judgment simply on the basis of the "pattern" presented. It is common for prosecutors as well as lawyers for defendants to try to show a "pattern of behavior" on the part of the accused that will support their plea. And it is often left up to the judge in sentencing trials to decide whether such a pattern has been demonstrated. And just as it is difficult to resist drawing conclusions on the basis of visual patterns, it is difficult to resist drawing conclusions on the basis of "patterns" of data, actions, and ideas.

Research on the matter of pattern recognition has not advanced as far as we would like. When does a set of numbers, actions, ideas constitute a pattern and when does it not? Can we assign numbers to indicate the degree of pattern in a set of data? What exactly is a pattern, anyway? Quantitative measures of patterns have been suggested and attempted but there is no well-established standard of evidence that goes beyond an appeal to our intuitions. Therefore, intuition remains a strong contender for the establishment of pattern recognition and thus coherence (see Hammond, 1996; Margolis, 1987, for further discussion).

Analysis

Analysis appears within the coherence metatheory when a person justifies his or her judgment by appealing to a set of logical rules. One set of rules that is often applied by researchers studying judgment and decision making under uncertainty are the rules of probability theory. Thus, for example, one may justify a prediction of the appearance of "heads" on the next throw of a die by appealing to the mathematics that probability theory entails. But if one does not have the advantage of training in mathematical probability—the competence required—then a quasi-rational or an intuitive approach will have to be taken (an approach that will occasionally produce a nearly correct answer). Mathematics, however, calls for exactness.

The question of how often and when incorrect intuitive judgments are made within a coherence metatheory of probability, and how badly incorrect they will be has been a topic of considerable interest (see chapter 5). One important consequence of this research endeavor was that it provided a new source of information to economists and philosophers who are still struggling with the applicability of the "rational man" theory. Some philosophers have accepted results of the Kahneman and Tversky studies as definite repudiation of the applicability of the rationality-based expected-utility model to human decision making (see, e.g., Horowitz, 1998). Although this debate is not likely to end soon, psychological research, particularly the work of Kahneman and Tversky, continues to gain attention from philosophers because it is directed toward the competence of analytical cognition within the coherence metatheory. (Philosophers appear to me to have lost interest in cognitive processes in relation to the correspondence metatheory.)

Not only philosophers and economists have given over the greater part of their attention to analytical cognition within the coherence metatheory but psychologists as well have elevated this aspect of cognition to the first rank in relation to measuring intelligence. For example, the Raven Progressive Matrices Test (Raven, 1962) is a highly visible and widely used test of intelligence that is based entirely on the proposition that what authors call "analytical intelligence" is critical to the ability "to reason and solve problems involving new information" (Carpenter, Just, & Shell, 1990, p. 404) in contrast to knowledge already acquired.

Thanks to the success of these researchers in modeling the cognitive processes of persons taking the test we now know that the test demands the successful discovery and use of five different rules. What determined the choice of these rules and the problems that exemplify them in the test? The explanation of the choice of these rules provided by Carpenter et al. is interesting in view of the remarks above regarding the need for an explicit theory of tasks. According to Carpenter et al.,

> John Raven constructed problems that focused on each of six different problem characteristics, which approximately correspond to the different types of rules that we describe later. He used his intuition and clinical experience to rank order the difficulty of the six problem types. Many years later, normative data from Forbes (1964) . . . became the basis for selecting problems for retention in newer versions of the test and for arranging the problems in order of increasing difficulty, without regard to any underlying processing theory. Thus, the version of the test [examined here] . . . is an amalgam of Raven's implicit theory of the components of reasoning ability and subsequent item selection and ordering done on an actuarial basis. (p. 408)

This is a fascinating account of the development of a test that has been used thousands of times, not only to measure the intelligence of a wide variety of people, but also as a marker for other tests. This account indicates that the original basis for the choice of the discovery of rules is defenseless; what should have been a coherent presentation of a series of ideas or concepts that provides the justification for why the discovery of rules is important is missing. Raven may well have had such a theory—and he may not—but it has never been available. Nor has a such theory been presented. The defense for the use of the test lies within the correspondence metatheory, that is, "it works." (Philosophers would label this "pragmatism.") The "actuarial basis" for item ordering and the various psychometric explorations of its relations with other tests (all equally likely to have dubious theoretical origins) provide the justification for its use for making decisions about the intelligence of people. And the modeling provided by Carpenter et al. of the cognitive activity induced by the problems affords, long after its development and wide application, further empirical evidence of the nature of the test. Thus, we see how a test that demands "analytical intelligence" from its subjects did not demand "analytical intelligence" from its originator—or at least did not demand that he make it explicit—nor from its developers, but did demand from them empirical competence, and thus correspondence competence.

Thinking about how this might have been otherwise helps to see the importance of coherence competence and the need for a theory of tasks. Instead of relying on "intuition" (whatever this may mean to Carpenter et al.) and "clinical experience" (whatever this may mean to anyone), Raven might have decided that he was interested in analytical competence, and then decided to discover what the logicians had to say about this, since analytical competence has generally been their field of interest. That might have led him, for example, to consult a book similar to R. Jeffrey's (1981) *Formal Logic*, where he would have found through a perusal of the table of contents that there are seven chapters, with the following headings: "Truth-Functional Inference; Truth Trees; Truth-Functional Equivalence; Conditionals; First-Order Logic; Undecidability; Completeness." That discovery might then have led him to decide that this book covered a reasonable variety of types of analytical challenges, and that he would be on safe ground if he were to include in his test of "analytical intelligence" a number of problems from each of these chapters. The content of his test then could at least be defended by pointing to his sampling method, rather than by an appeal to his "intuition" and "clinical experience." In the empirical investigations that would follow, the investigators would then have the benefit of knowing what they were investigating.

Intuition and Analysis Within the Correspondence Metatheory

Within the framework of the correspondence metatheory, intuition can be readily recognized as a mode of cognition. For example, when we make visual judgments of size and distance and other aspects of our physical surroundings, we do so intuitively. It is that kind of correspondence competence that we are familiar with, so familiar that we do not take note of it, yet it remains something of a mystery and certainly something of a miracle. Think of your ability to drive an automobile down the road at 60, 70, or 80 miles an hour amidst a dense crowd of other drivers doing the same without incident. Or the ability of airplane pilots to keep within a tight formation, wing tips only a few feet apart, while flying at 600 or 700 miles per hour, and successfully conducting maneuvers involving sharp vertical and horizontal changes. Although examples of marvels of perception are easy to come by, their explanation is not. No one can explain or describe to themselves the mental operations that make such empirical accuracy possible; indeed, the process is so effortless, no one tries. Even the explanations by scientists still leave much to be desired. (The matter of correspondence and coherence competence is discussed in more detail in chapter 3.)

As we move away from our extraordinary capacity for empirical accuracy in the physical world, generally based on multiple fallible indicators, our competence decreases progressively. Generally, as the task materials change from physical objects and events to social objects and events our competence decreases. That is, we are not nearly as competent in our judgments of social objects, people, as we are in our judgments of physical objects. This is not because *we* change in our abilities to make use of multiple fallible indicators, but because in person perception the indicators we must rely on (facial expressions, body language) are more ephemeral, more difficult for us to measure, for the indicators themselves require perceptual judgment—and the *social task system* is far more uncertain (facial expressions don't always mean the same thing) than most physical task systems. All of these generalizations have been demonstrated in the laboratory and elsewhere with a variety of materials, and most important, all this can be shown to be mathematically true. In short, these theoretical expressions have been shown to be *coherent* (mathematically true), as well as to be empirically true, and thus to have achieved *correspondence* as well (see Cooksey, 1996).

We can also take an analytical approach within the correspondence metatheory. If, for example, we are dissatisfied with our intuitive ability to predict the weather, say, we can use statistical methods to do so, and of course, that is what weather forecasters often do. Similarly with conceptual processes, various statistical methods may be applied to data that one must organize into a judgment and thus produce an analytically based judgment whose goal is accuracy of prediction. Unhappily, there are many situations in which our correspondence competence is less than what we would like it to be—in our judgments of character, say—but there are no analytical models available that will help us (see Cooksey, 1996; Hammond, 1996, for examples).

In short, both metathories can accommodate both forms of cognition. Recognition of this fact is of considerable importance for the study of the disruption of stabilized relations, a matter developed in chapters 8 and 9. We turn first to a theory of judgment that builds on this discussion of intuition and analysis within the two metatheories.

Cognitive Continuum Theory

Intuition and analysis are theoretical terms that are used by all researchers in the field of judgment and decision making. That means that, as I have shown above, they are used by those working within the coherence metatheory, as well as by those working in the correspondence metatheory. These terms also appear frequently in ordinary discourse, even in the newspapers, but with a different implication than is intended here. In ordinary discourse, and scientific discourse as well, the implication is that the terms intuition and analysis are mutually exclusive. This implies that there is a *dichotomous* relationship between them; if a person changes from intuitive cognition to analytical cognition, it is assumed that they are switching completely from one form of cognition to the other. That is not the position taken here, however. I argue that there is a continuous relationship between the two, that there is a *continuum* between intuition and analysis, thus the term *cognitive continuum theory*. And from the dismissal of the assumption of a dichotomous relationship between intuition and analysis and its replacement with that of a continuum, a number of contingencies follow. I express these in terms of five premises (see Hammond, 1996, pp. 147–202, for a detailed explanation).

Premise 1: A Cognitive Continuum

Various modes, or forms, of cognition can be ordered in relation to one another on a *continuum* that is identified by intuitive cognition at one pole and analytical cognition at the other.

The idea of intuition and analysis lying at poles of a continuum is a new one. Traditionally, they have been thought of as two distinct types of cognitive activity separated by a sharp boundary; they were thought to have an either-or relationship. This conception left no room for a form of cognition that included elements of both intuition and analysis. The above premise does provide room for the idea that various forms of cognition do include elements of both, and that these can be ordered in terms of the number of elements of either that are included. Thus, some forms of cognition may include a larger number of intuitive elements of cognition than analytical ones, and thus be located closer to the intuitive pole of the cognitive continuum. And, of course, some forms may contain a larger number of analytical elements and thus be located closer to the analytical pole. For example, judgments of another person's suitability for employment may be based on a rapid judgment of the person's likability, a guess at their level of intelligence, and a review of their work history. The first two elements are largely intuitive, the third requires some analytical work; did the person perform the kind of tasks she or he will be expected to perform in the new situation? Only the latter element will require analytical cognition. As a result, this judgment lies closer to the intuitive pole than the analytical pole. Similarly, with respect to analytical judgments, they may be almost wholly analytical but not quite. For example, in making a calculation of how long an automobile trip will take, one might engage in the analytical work of looking up the number of miles to be traveled, count how many miles will be on a freeway versus how many will be on side roads, but then have to make a guess of the average speed that will be maintained on each. In this case, as in so many, one is as analytical as one can be and as intuitive as one must be. This judgment will lie closer to the analytical pole of the cognitive continuum than the intuitive pole. Making this distinction allows us to think in a more differentiated manner about cognitive activity, but imposes the burden of demanding that we somehow specify exactly *where* on the cognitive continuum any given judgment will lie, a matter I take up below.

Implications for the Study of Disruption of Constancy

On the traditional assumption of a dichotomous relationship between intuition and analysis, we would anticipate that the effect of disruption would be to shift cognition from one of these forms of cognition to the other, completely and decisively, in an all or none fashion. But if we assume a continuous relationship between these two forms of cognition, we then have an opportunity to ask new questions: For example, *how far* does disruption move cognitive activity on the cognitive continuum from one location to another? That is, does a minor disruption move cognition from analytical processes to intuitive processes only slightly? Does a greater disruption make a person *significantly* less analytical and *significantly* more intuitive, thus making their judgments less than fully defensible? That change would be particularly important for operators trying to defend judgments made during the disruption, trying to explain why they decided to do what they did. Also we can now ask: What, exactly, determines the amount of movement? Is it the *amount* of disruption, or the *kind* of disruption that has the greatest effect? Is movement on the cognitive continuum a linear or nonlinear function of disruption? These questions have never been addressed because the idea of a cognitive continuum is a new one. The answers are likely to be far more informative, however, than simply asking such general questions as "does noise impair judgment," without specifying the kind of disruption noise might create in which sort of task, or specifying the kind of judgment that might be affected.

The use of the concept of a cognitive continuum rather than a cognitive dichotomy also allows us to address the effects of disruptive conditions on persons at the highest level of decision making in government. The publication of the material secretly recorded on White House tapes by President Kennedy, particularly those made during the Cuban missile crisis, provides us with an opportunity to do so (see May & Zelikow, 1997). Examination of that material from the point of view of a cognitive continuum is informative in a way that historical or journalistic efforts are not. For example, in the following paragraph May and Zelikow show how Kennedy relies on intuitive cognition to a large degree in his decision to go ahead with the Bay of Pigs invasion of Cuba. They state that

> afterward, he recognized that he had not only listened to too few advisers but that he had given the issues too little time [thus suggesting to us that his cognitive processes were closer to the intuitive pole]. Eisenhower, . . . subjected him to a staff school quiz. "Mr. President," Eisenhower asked, "before you approved this plan did you have everybody in front of you debating the thing so that you could get the pros and cons yourself and then made the decision . . . ?" Kennedy had to confess that he had not. (p. 28)

In other words, Eisenhower showed him that he had not been as analytical as he should have been. But after the bitter experience of the failed invasion, and as a confrontation with the Soviets was building up, May and Zelikow indicate that Kennedy began "applying lessons from the Bay of Pigs affair." Now he "searched widely for advice, insisted on going over and over alternative courses of action, and pressed for imagination to expand the menu" (p. 30), thus moving the cognitive activity of all concerned toward the analytical pole.

The poles of the cognitive continuum are seldom reached, however. In the following very important exchange, Kennedy and his advisors try to figure out how to respond to two very different messages they have received from the Soviets regarding their placement of missiles in Cuba. Note the rambling nature of the discussion; it is far from an analytical examination of the situation, yet never goes so far as to be based wholly on intuition.

> PRESIDENT KENNEDY: Well now, what we have to do first is get, I would think, very quickly to get a chance to think a little more about it. [Note that the president is under time pressure to produce analytical cognition; he needs to develop a defensible reaction to the confusing letters; thus he "very quickly" wants to "think a little more about it."]
>
> But what we ought to say is that we have had several publicly and privately different proposals, or differing proposals from the Soviet Union. They are all about complicated matters. They all involve some discussion to get their true meaning. [Notice that even when the future of civilization is at stake the heads of state do not make the meaning of their communications clear to one another; Kennedy is trying to figure out the "true meaning" of the messages from the Soviet Union; he isn't even sure of their authorship. Analytical cognition is virtually precluded by these circumstances.] We cannot permit ourselves to be impaled on a long negotiating hook while the work goes on on these bases. I therefore suggest that the United Nations immediately, with the cooperation of the Soviet Union, take steps to provide for cessation of the work [installation of the nuclear missiles in Cuba], and then we can talk about all these matters, which are very complicated.
>
> BUNDY: I think it will be very important to say at least that the current threat to peace is not in Turkey; it is in Cuba. There's no pain in saying that, even if you're going to make a trade later on.
>
> I think also that we ought to say that we have an immediate threat. What is going on in Cuba: that is what has got to stop. Then I think

we *should* say that the public Soviet, the broadcast, message is at variance with other proposals which have been put forward within the last 12 hours. We could surface those for background.

PRESIDENT KENNEDY: That being so, until we find out what is really being suggested and what can really be discussed, we have to get something in words. Maybe we can see if the work's going on.

BUNDY: That's right.

PRESIDENT KENNEDY: There isn't any doubt. Let's not kid ourselves. They've got a very good proposal, which is the reason they made it public—

BUNDY: What's going on, while you were out of the room, Mr. President, we reached an informal consensus that—I don't know whether Tommy agrees—that this last night's message was Khrushchev's. And this one is his own hard-nosed people overruling him, this public one. They didn't like what he said to you last night. Nor would I, if I were a Soviet hard-nose.

THOMPSON: I think the view is, the Kreisky speech, they may have thought this was our underground way of suggesting this, and they felt that—

UNIDENTIFIED: Who said this?

PRESIDENT KENNEDY: The only thing is, Tommy, why wouldn't they say it privately if they were serious? The fact that they gave it to me publicly, I think they know the complexities. (May & Zelikow, 1997, pp. 512–513)

In such conversations, certainly serious ones, in which intelligent, experienced professional people are carefully monitoring every word uttered by everyone else, we can see that although the arguments *attempt* to meet the standards of analytical rationality—they are not irresponsible people—they seldom do. Nevertheless, they are seldom challenged; we seldom see remarks criticizing a speaker for being irrational or inconsistent or illogical, even in these extreme circumstances where the stakes are so high as to preclude mere politeness. What we see are attempts to be as analytical as possible within a context in which it is impossible to be fully analytical. In short, we see a context where the mode of discourse is quasi-rational. Full analysis is impossible, and full intuition would not be acceptable.

The participants are aware of the desirability of rationality, however. After what appears to be an emotional outburst by Robert Kennedy, Robert McNamara, apparently wishing to move the conversation toward a more rational analytical form, said:

Mr. President, this is why I think tonight we ought to put on paper the alternative plans and the probable, possible consequences thereof, in a way that State and Defense could agree on. Even if we disagree, then put in both views. Because the consequences of these actions have not been thought through clearly. The one the Attorney General just mentioned is illustrative of that. (p. 99)

He wants an opportunity to "think through" the "alternative plans and the probable, possible consequences thereof," a process he apparently believes has not yet been done.

Thus, the conversations recorded by Kennedy provide an excellent document for examining the cognitive processes of people exchanging thoughts under natural circumstances. It illustrates the nature of the cognitive activity that guides the discourse of intelligent, skillful, experienced professionals. Moreover, these professionals are operating under what the layperson would call "stress," but what is here called a threat to ***disruption of stabilized relations***, between two powerful entities on the brink of destroying a large fraction of civilization that includes themselves. It would be difficult indeed to create a laboratory experiment that would reproduce these circumstances.

Multiple Fallible Indicators as "Signals" in International Communication

Earlier I described the persistent role of multiple fallible indicators in judgments by early hominids to find food and identify enemies, as well as their use in navigation by animals, birds, and insects. And I described their use in a wide variety of contexts by modern humans, as in economics, sociology, and elsewhere (see Hammond, 1996, for a detailed treatment). The same use of indicators is often described by political scientists and journalists in international communication as the "sending of signals," which are, in fact, multiple fallible indicators. One of the most important uses of "signals" occurred during the Cuban missile crisis, during which the Russians and the Americans were attempting to indicate their intentions to one another with a combination of words and actions, none of which were unambiguous. Indeed, these "signals" can sometimes be quite subtle. May and Zelikow, for example, describe Kennedy's signaling to Moscow in this way: "In his speeches and in diplomatic communications, Kennedy was careful to speak of West Berlin and not, as had been customary earlier, simply of Berlin. It sent a calculated signal that the United States would not necessarily feel compelled to act if the Soviets or East Germans were to do something in the portion of Berlin already under their control" (p. 31). Did the Soviets see this signal? If so, did they interpret it correctly (i.e., as Kennedy intended)? How did it fit with the many other signals that were being sent? Since history is replete with examples of misunderstandings of intentions conveyed by such signals, one must wonder if this is the way international communication should be conducted.

Premise 2: Common Sense

The forms of cognition that lie on the continuum between intuition and analysis include elements of *both* intuition and analysis and are

included under the term *quasi rationality*. This form of cognition is known to the layperson as common sense.
"Common sense" is a phrase that scientists, psychologists in particular, try to avoid; it is part of the vernacular, part of our everyday language that has lost any specific meaning it might have had. Dictionaries are not helpful, for they generally tell us that common sense implies "good judgment," "wisdom," "prudence," and even refer to such colloquial terms as "horse sense." Although I will not use the words "common sense," I believe it is of great importance to acknowledge the fact that it refers to a cognitive activity of great importance—namely, quasi-rational cognition.

Quasi-rational cognition is a term that is required by virtue of Premise 1. Specification of a cognitive continuum between analysis and intuition demands a place for a form of cognition that lies between analysis and intuition and includes elements of both. That is important to us because it is generally believed that "stress" deprives us of our common sense, and that suggests that disruption of a person's stabilized relations with the environment causes cognition to move from the center of the cognitive continuum in the direction of either pole. That hypothesis is examined in what follows.

First, however, we have to come to grips with the matter of deciding exactly *where* on the cognitive continuum *any* act of judgment lies. How do we know, for example, whether this particular judgment was *more* intuitive, or *more* analytical, than any other one? To answer that question, we need to put forward a specific assignment of properties to both intuition and analysis, and to indicate a procedure for measuring them, a matter to which we now turn.

Determining the Location of a Judgment on the Cognitive Continuum

The first step determining the location of a judgment on the cognitive continuum is to assign properties to intuition and analysis; that is done in Table 7.1 Elsewhere (Hammond, 1996, p. 181) I have used these properties to describe analytical cognition as follows:

> A high degree of cognitive control . . . implies a tight command over cognitive activity, and that requires a high degree of awareness of that activity (knowing what you are thinking) which in turn implies that there will be a fixation on just a few indicators rather than a wide and interchangeable use of indicators (consistent with low cognitive control). Additionally, the rate at which cognitive activity takes place in analytical work is slow (it usually involves checking for mistakes), in contrast to rapid intuitive judgments that occur in seconds. Judgments often depend on memory, but analytically-based judgment depends on recalling principles related to the task (e.g., the inertial principle in physical circumstances), in contrast to the recollection of specific, often

Table 7.1 Properties of Intuition and Analysis

	Intuition	Analysis
Cognitive control	Low	High
Awareness of cognitive activity	Low	High
Amount of shift across indicators	High	Low
Speed of cognitive activity	High	Low
Memory	Raw data or events shared	Complex principles stored
Metaphors used	Pictorial	Verbal, quantitative

vivid ones, in intuitive cognition. Both forms of cognition rely on metaphors: Analytical cognition tends to rely on verbal or quantitative ones, whereas intuition will rely on pictorial, easy-to-use metaphors. (pp. 181–182)

In a study of highway engineers, R. Hamm, J. Grassia, and T. Pearson, and I (1987) measured the extent to which each of these properties of a person's cognitive system was apparent in his judgments of the safety, capacity, and aesthetics of various highways. Certain properties of intuition and analysis in Table 7.1 were applied, depending on procedural practicality. For example, cognitive control, type of organizing principle, error distribution, and differential confidence were determined. The measure of each property was aggregated by means of a simple sum, thus providing a measure for the Cognitive Continuum Index (CCI); the resultant number indicated the location of each person's judgments on the cognitive continuum (see Hammond, Hamm, Grassia, & Pearson, 1997, pp. 156–157, for details of the measurement procedure).

Locating a person's cognitive activity on the CCI is of critical importance, for a quantitative indicator of that location makes it possible to compare one person's location with another's and thus to compare one person's cognitive activity with anothers, as well as to measure the amount of shift in a person's cognitive activity produced by different task circumstances (such as disruption of stabilized relations). That is, use of the CCI makes it possible to compare the amount of quasi rationality—the nearness to one pole or the other on the continuum—exhibited by one person with that of another, and thus to describe one person as more analytical (or intuitive) in those circumstances. Use of the CCI also makes it possible to show that a disruption in stabilized relations resulted in a shift in a person's cognitive activity from being highly analytical to being highly intuitive. I make use of this theory below (chapter 10) to describe the cognitive activity of both Commander Rinn and Captain Haynes, and thus to indicate that its use is not limited to the laboratory.

Implications for the Study of Disruption of Constancy

If common sense—quasi rationality—occupies the middle ground between intuition and analysis on the cognitive continuum, it would be natural to ask: Does disruption deprive us of common sense? Does it drive us to judgments that we would not ordinarily make? The layperson's answer to these questions will be "yes," of course. More and more frequently the explanation for mistaken judgments is: "I was under stress" But the explanation based on Cognitive Continuum Theory will require more details. It will require knowledge of the initial point on the cognitive continuum, at which the person's cognitive activity was located when the "stress," that is, the disruption, occurred. If the person's cognitive activity was located at the analytical pole, that is, was engaged in analytical cognition, and thus using a rule to make a decision when the disruption occurred, then it would be reasonable to ask: Did the disruption drive the person *toward* the middle ground of the cognitive continuum? If the person was a trained operator of a complex process, we should expect that disruption (any unexpected event for which the operator was not trained) would drive the operator *toward,* not away from, common sense; if change is expected, that is the only change that can occur. So, contrary to folklore, disruption of analytical cognition can drive a person to common sense.

Can a person be driven *away* from common sense by disruptions? Yes, in either direction. If the person is operating at the midpoint of the cognitive continuum, that is, operating with the use of quasi-rational cognition at the time disruption occurs, they will be driven to the intuitive pole of the cognitive continuum if there is time pressure, a demand for a quick decision. For immediacy is a prime condition that induces intuition. But if the unexpected event cannot be understood by the use of quasi-rational cognition, and there is time to think, to reason, to seek a coherent explanation for why the unexpected event may have occurred, then the person will be driven away from common sense toward analytical cognition. The matter of time has always been recognized as important, but its role in judgment has not been appropriately recognized because of lack of recognition of a cognitive continuum.

It is interesting that nearly 100 years ago one of the Wright brothers thought about this matter in connection with flying their airplane. In a lecture to the Western Society of Engineers on June 24, 1903, Wilbur said: "Before trying to rise to any dangerous height a man ought to know that in an emergency his mind and muscles will work by instinct rather than by conscious effort. There is no time to think" (Wescott and Degen, 1983, p. 83).

The CCI, described immediately above, provides a means of quantifying the direction and degree of movement on the cognitive continuum and thus makes it possible to predict the cognitive activity of the person making the judgments. Therein lies an important advantage of cognitive continuum theory; it allows us to think about cognition in a more explicit, differentiated way than is afforded by folklore, and it provides testable predictions and a means for testing them. But it may be that the most important comparison to be made is between the location of the person's cognitive activity on the CCI and the location of the task on its index, to which we now turn.

Premise 3

Cognitive tasks can be ordered on a continuum with regard to their capacity to induce intuition, quasi rationality, or analytical cognition. We have always had rich theories of behavior, rich theories of personality, and rich theories of cognition. By "rich" I mean that we have had numerous concepts and numerous linkages among them to be used for the description of such phenomena. But, as I have indicated earlier, we have almost no theories of environmental circumstances, no theories that will provide us with a description of a specific set of circumstances that can be related to and differentiated from another set of circumstances, and therefore no systematic means of generalizing our findings over tasks. This omission has been particularly damaging in the field of judgment and decision making in general, and in the field of "stress" in particular.

Premise 3, however, makes it imperative that a theory of tasks be made specific, for Premise 3 asserts that tasks can be *ordered* on the cognitive continuum. That means that there must be a procedure for assigning properties to each judgment task, and that these properties be somehow assigned a quantitative value that allows a specific task to be placed on a continuum in ordered relation to other tasks similarly treated.

Establishing a Task Continuum Index: An Example

In the study of highway engineers mentioned above, a location on the Cognitive Continuum Index (CCI) was determined for each engineer's cognitive activity based on his judgments of highway safety. The next step was to develop a measure of the location of each judgment task on a Task Continuum Index (TCI). This was done so that a quantitative comparison could be made between the properties of a task and the properties of cognitive activity. We wanted to answer the question of whether a task's location on a Task Continuum Index reliably induced a specific form of cognitive activity.

To accomplish this step, the concepts of the Cognitive Continuum Theory were applied to both the surface (the display layer) and the depth (the inference layer) of the task. That is, at the surface layer, the information was displayed in three ways: either to induce intuitive cognition (pictures), quasi-rational cognition (bar graphs), or analytical cognition (numbers). At the depth layer, the variable to be inferred from the surface

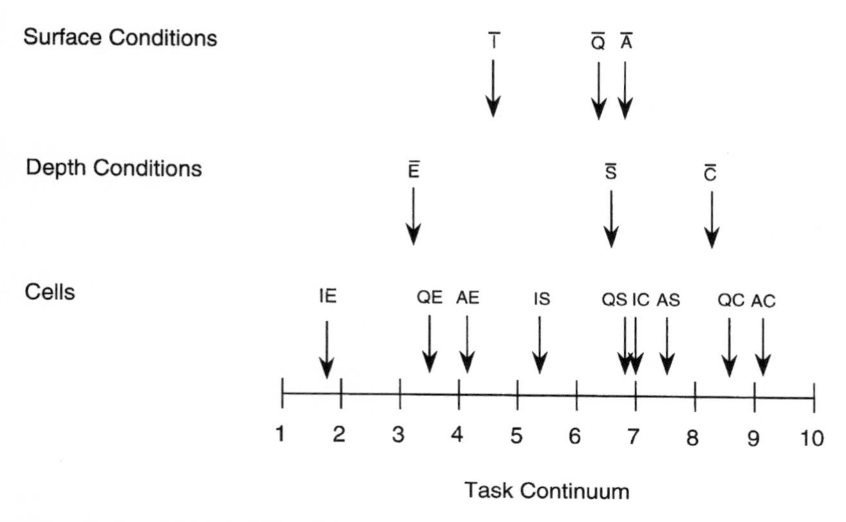

Figure 7.1. Ordering tasks on the Task Continuum Index.

data was arranged to be one that would induce either intuition (aesthetics of highways), quasi rationality (the safety of highways), or analysis (the carrying capacity of highways). These conditions were crossed, thus producing nine judgment tasks, the distribution of which can be seen in Figure 7.1.

The specific location of each display and each depth variable on the Task Continuum Index was determined by measuring their properties as defined in terms of the Cognitive Continuum Theory. This is not the place to discuss the details of this procedure (they can be found in Hammond et al., 1987; also 1997). The significant point is that Figure 7.1 shows that it is possible to order cognitive activity on a continuum from analysis to intuition and that it is possible to order judgment tasks on a parallel continuum as well. Once this is done, a study is securely anchored. We know what the task conditions are (they are specified in theoretical and empirical terms), and we also know what the cognitive conditions are (they are also specified). At this point, therefore, we are prepared to make predictions and to indicate the range of generality intended, over both task and cognitive conditions. It will no longer be necessary to make open-ended generalizations of the nature of what people do or do not do based on a few studies that specify neither task properties nor cognitive properties and thus offer no limits to generalizations (Smith & Kida, 1991).

A Task Continuum Index Within the Coherence Metatheory?

The CCI and TCI were developed in connection with research carried out within the correspondence metatheory. Can a CCI and TCI be developed within the coherence metatheory? They certainly should be. (Remember that both intuition and analysis can be carried out within the coherence metatheory as well as the correspondence metatheory.)

This is not the place to provide a detailed answer to this question but it should be answered. My view is that the first hypothesis to be tested should be that the cognitive continuum as described above will be appropriate. That is, the list of task properties and cognitive properties will be found to be applicable, although the organizing principle may have to be changed, and more than one needs to be considered. It will be especially important to undertake this task in connection with the problem of disruption.

Implications for the Study of Disruption of Constancy

Premise 3 and its explication assert that it is task conditions that induce changes in cognition. Premise 3 also postulates that cognition *covaries* with task conditions, that is, location of cognition on the CCI changes as the location of the task changes on the TCI. Thus, for example, if there is a change in task properties (e.g., a sharp reduction in time allowed for a decision) such that the location of the *task* on the TCI moves sharply toward the intuition pole, then we should anticipate that *cognition* will become more intuitive. And, of course, vice versa; changes in task conditions that signify that the task allows more time for cognition and therefore has moved toward the analytical pole should lead us to anticipate an increase in analytical cognition.

This prediction carries implications for the study of disruption of constancy because it is precisely those unexpected, often inexplicable, task events that demand immediate decisions that drive cognition away from analysis toward the intuitive pole of the cognitive continuum. Similarly, task circumstances that ordinarily require no more than quasi-rational cognition may change in a way that demands that more thought be given to them. If more time or more rule-use is demanded, then a move toward analytical cognition will be induced by these new task conditions (if time permits). Events that demand a move toward analytical cognition may be even more disruptive than those demanding a move toward intuition since intuition needs no models, no guidance—it does what it does—whereas analysis must always find a defensible solution—usually a defensible organizing principle, often an ideology—capable of being retraced. In the application of Cognitive Continuum Theory to the disruption of constancy, it is the specifics of the disruption—the change in the location of the task on the TCI—that allows us to predict change in cognitive behavior.

The premises of the theory, namely, that cognitive behavior changes as a function of task conditions, together with the TCI (a means for measuring task changes) and the CCI (a means for measuring cognitive change), thus provide a basis for predicting the effects of disruption of constancy on human judgment. Implicit in the theory, then, is the argument that what the layperson terms "stress" arises out of unanticipated changes in task conditions. We could see those changes in Captain Haynes's cognitive activity. In the next section, however, we will consider *movement*, and thus *oscillation* on the cognitive continuum due to changes in task conditions.

Premise 4: Dynamic Cognition

Cognitive activities may move along the intuitive-analytical continuum over time; as they do so, the relative contributions of intuitive and analytical components to quasi rationality will change. Successful cognition that maintains constancy with a stable environment inhibits movement; failure and loss of constancy with a stable environment stimulates movement on the continuum. Maintaining constancy with a changing environment may require oscillation on the cognitive continuum.

An example of oscillation between intuition and analysis due to an endogenous task system failure is provided by the two-man crew of a 747 cargo jet that took off from New York at 2:40 A.M. bound for Tokyo on December 12, 1991 (Carley, 1993, p. A1). According to Captain William Jackson, about three hours later (5:20 A.M.) the autopilot system began to malfunction—although the crew was unaware of this. Very soon (within 15 seconds) the plane rolled to a 30-degree bank, at which point warning lights lit up indicating failure of the inertial guidance system. Subsequent events were described by Carley as follows: "After exclaiming 'What's going on?' Captain Jackson looked at his attitude indicator, an instrument that provides an artificial horizon. To his amazement, the whole thing was black. . . . 'We've lost instruments,' the captain exclaimed" (p. A6). The copilot looked at an auxiliary display that was working and saw that the plane was now in a 90-degree (!) bank. At that point the copilot thought the *instruments* had failed because his senses told him otherwise: "I couldn't believe we were rolled over on a knife edge" (p. A6). (It should be explained that the plane was flying at night; there were no lights on the ground. That is, the plane was in a black envelope. Under these conditions proper orientation—correspondence constancy—can be maintained only by an instrument known as an "artificial horizon.") The copilot then decided the *instruments were correct* and that his kinesthetic senses were wrong. He then grabbed the yoke and attempted to correct the 90-degree bank. But the plane didn't respond immediately, and he returned to a belief in his kinesthetic senses, turned the yoke in the opposite—exactly wrong—direction before deciding *once more* that the instruments were *right*. He then correctly returned the plane to the proper orientation. Thus, we see intuitive (kinesthetically produced) judgments oscillating with (analytically based) judgments based on technological instruments in an effort to restore correspondence constancy that was disrupted by a loss of coherence constancy.

But the plane was still diving, its speed was accelerating and it began to vibrate loudly. The copilot recalls: "I thought, 'My God, the plane's going to fall apart.' . . . I wanted to pull the plane out of that dive to reduce speed. But I *thought* [italics added], if I pull back too hard [tilting the elevators on the tail], wind forces are going to rip the tail right off" (p. A6).

To see the oscillation between intuition and analysis note that, first, the copilot analytically tries to right the airplane, that is, he responds to what an instrument tells him to do; his kinesthetic senses, however, contradict the instrument information and he makes exactly the wrong response, not an uncommon event in flying an airplane when one cannot see the horizon, and the cause of many fatalities prior to the invention of the artificial horizon. Then as a result of looking at the instruments he sees what he has done and corrects his mistake. He is oscillating between intuition (responding to misleading sensory experience mediated by receptors of which the person is unaware) and analytical cognition (responding to information from an instrument that he has been trained to use) and thus correcting a certainly fatal intuitive error. Moreover, the copilot *continues* to be analytical; he *thinks* in terms of an if-then proposition: "*If* I pull back too hard [then], wind forces are going to rip the tail right off." Thus, we see that he can *retrace* his cognitive activity, something that cannot be done following intuitive cognitive activity (recall Captain Haynes's insistence that his airplane could make only right turns). But the copilot and the captain did not "pull back too hard," and despite considerable vibration and holes in the wing, the crew managed to get the airplane under control and landed it safely at an airport an hour and 20 minutes away. These events illustrate how cognition can oscillate between intuitive and analytical cognition.

These events also indicate how an endogenous disruption of coherence competence (failure of instruments) can lead to a disruption of correspondence competence (loss of spatial orientation) and how these events can lead to oscillation between intuition and analysis.

Implications for the Study of Disruption of Constancy

As yet we have no information on what constitutes the most effective rate of oscillation for various circumstances (see Hammond, Frederick, Robillard, & Victor, 1989, for an effort to study this phenomenon). It is easy to imagine, however, that in circumstances that are not benign, time will seldom permit oscillation (movement on the cognitive continuum) to be orderly. Severe demands for remedial action may result in cognition oscillating so rapidly between the poles of the continuum that the process becomes chaotic. Or the rate of oscillation may be reduced or the process truncated so that cognition becomes more closely associated with one mode than it should. Although the dynamic character of cognition has hardly been touched by judgment and decision researchers (but see Brehmer, 1996), it may well be that the study of the effects of disruption of constancy will encourage investigation of movement on the cognitive continuum. Such movement may be an important feature of leadership under conditions of disruption, and may require a new form of training for those who are expected to be followers. (I will provide an example in chapter 10). In short, we don't know whether disruption of constancy results in greater or less oscillation, nor do we know whether cognition can keep up with rapid changes in task conditions that induce movement on the cognitive continuum. But the examples I will provide in the next chapter will indicate that change is highly dependent on whether disruption is endogenous or exogenous.

Premise 5: Pattern Recognition and Functional Relations

Human cognition is capable of both pattern recognition and the use of functional relations.

These two concepts—pattern recognition and functional relations—occupied the attention of psychologists throughout much, if not most, of the first part of the twentieth century. The question was: Were our cognitive processes controlled by one or the other? The struggle over this question was hard and has been recounted often. Generally, the battle seems to have been lost by the Gestalt psychologists, the champions of the struggle for the dominance of pattern recognition (for an exception see Pennington & Hastie, 1991, 1993). Although the issue is no longer framed in exactly this way in cognitive psychology, it can hardly be dismissed in the field of judgment and decision making.

But academic students of judgment and decision making have apparently found no use for the concept of pattern recognition. The only author to insist on the importance of pattern recognition as a controlling feature of human judgment is H. Margolis (1987), whose book was generally ignored by judgment and decision researchers. No contemporary book in the field of judgment and decision making discusses the concept of pattern recognition other than to denigrate it as a source of error (for an example, see Hogarth, 1987; see also Hammond, 1996, pp. 196–201, for a full discussion). There is good reason for this. In the beginning of the work on this topic it was assumed, as all laypersons assume, that pattern recognition played a large role in the judgment process. But after roughly 20 years of research on the topic, they would argue that there was no empirical basis for this belief (see, e.g., Hammond, McClelland, & Mumpower, 1980; for a more recent treatment, see Yates, 1990). Models of judgment that included only linear relations between cues and judgments offered good evidence that patterns were not employed in the judgment process, despite reports by the subjects that they were. Nevertheless, the belief in pattern recognition remains as strong as ever among laypersons and nonpsychologists today as it ever was. The question now is: When is pattern recognition employed and when is it not? Strangely, this question seems to be of little or no interest to students of judgment and decision making.

The reason for this strange state of affairs—researchers denying the significance of the concept of pattern recognition—lies in the problem addressed in the discussion of Premise 4; a lack of a theory of tasks. It seems obvious that certain tasks—they are easy to construct or identify—will evoke cognitive activity that is based on pattern recognition and other events. All those Gestalt figures in the elementary psychology textbooks are there for just that reason; they are patterns that evoke pattern recognition immediately and easily; there is no denying them. Similar patterns can be seen on the nightly weather reports on the television set when the meteorologist shows the movement of a frontal system. What the viewer never sees, is what the weather forecaster at the weather station computer sees—a picture of functional relations, such as a plot of the relation between temperature and altitude, often on the same screen that shows the meteorologist the pattern of the frontal movement. Thus, the weather forecaster's predictions are based on what the judgment and decision researcher ignores—an *alternation* between pattern recognition and the use of functional relations. For that is what the forecaster does; she or he alternates between the use of patterns (the frontal system) and functional relations between physical variables. And, no doubt, other users of computer displays often do the same. What cries out for research is the question of the form of cognitive activity evoked by this dual presentation of patterns and indicators. And, of course, we need to know how destabilization affects this cognitive activity. In other circumstances not so rife with uncertainty, it takes only *one* deviant data point to reject the "pattern of behavior."

Patterns and Functional Relations: Coherence and Correspondence

Patterns do not stand outside our metatheories of coherence and correspondence. Indeed, one of the unquestioned achievements of Gestalt psychology was to show us the strength of our intuitive predilection for the *perception* of coherence, even when it is not quite there. This was done by demonstrating that we see "wholes," that is, complete patterns (circles, rectangles, and other geometrical figures) when, in fact, the stimulus figure is slightly incomplete (and briefly presented). As a result, the implicit and explicit perception of patterns is widely believed to be a highly compelling and significant part of our cognitive apparatus. Readers may ask themselves how often they have heard someone say "there is a pattern of activity here that indicates . . ." But patterns must meet the test of coherence. For example, when a weather forecaster points out a pattern of weather activity and then concludes a severe storm is on the way, he or she may be challenged by someone who sees an imperfection in pattern; a data point that does not fit the pattern. Whether the challenge is sustained will depend on the centrality of the data point to the depiction of the pattern. Similarly, with the description of a person's behavior, faced with a myriad of items of information about a person convicted of a crime, judges often justify a sentence by referring to a "pattern of activity" of past behavior.

Consequently, discussions of the perception of patterns bring us to a discussion of coherence. And that means that the meaning of coherence needs to be broadened to include not only the demand for consistency that is exhibited in mathematics and logic but also the consistency of any set of data, whether this be data about a weather system, a person, or any system. Thus, when we look for a "pattern" in behavior, we are asking about the consistency, or coherence, of that behavior. Similarly, with respect to a person's religious or political beliefs, or someone's belief system in general, or their behavior in general. In short, consistency, or coherence, is not a criterion that can be applied only to mathematical or logical systems; we can look for coherence or its absence in any system. And as every text on memory will remind us, coherence—and incoherence—play a strong role in what we remember, both with respect to what we expected to happen but didn't, as well as what we didn't expect to happen but did.

Implications for the Study of Disruption of Constancy

So far as I know, we have no information on the effects of disruption of constancy on pattern recognition or the use of patterns when recognized. Nor do we have information on the effects of the commonly used "stressors" on the disruption of coherence constancy. There are studies of the effects of these "stressors" on a variety of judgments but no systematic studies.

In discussing Premise 4 considerable emphasis was placed on the topic of *oscillation* between intuition and analysis. Here, in Premise 5, we focus on the *alternation* between the use of pattern recognition and the use of functional relations in reaching a judgment under conditions of disruption. It is important to note that whereas oscillation implies a *continuous* process, alternation implies a *discontinuous*, all-or-none, process. As in the case of oscillation, little or nothing is known about the most *effective* rate of alternation under benign *or* disruptive conditions. Is it better to alternate rapidly or slowly between the use of the coherence theory of truth and the correspondence theory of truth? Should alternation take place at all? We don't know the answer to either of these questions. But while both oscillation and alternation may be normal in normal situations—we seem to have learned to live with it—oscillation and alternation can become critical in crisis situations. As we saw in our analysis of the Kennedy tapes of the discussions of the Cuban missile crisis, it can lead to confusion in the dialog and failure to understand how other persons arrived at their judgments. This in itself may not produce catastrophe, but it leaves the outcome more susceptible to chance. If the participants do not understand the difference in these two ways of making judgments, misunderstanding is almost inevitable, and of course, occasionally fatal. As matters stand now, participants will not understand the differences in these two ways of making judgments.

The Significance of the Cognitive Continuum Index and the Task Continuum Index

Methodological Significance

The CCI and TCI described above are merely first steps toward developing a concrete theory of human judgment. Primitive as they may be, the purpose of presenting them here is to make clear that they are *necessary* steps. They are necessary because they are quantifiable steps toward a broad theory that specifies the essential characteristics of both person and environment, as well as their interaction. No other theory does that. Without this broad scope that denotes both person and environment, we remain condemned to a constant discovery of "effects" (e.g., the "endowment effect," etc.), the generality of which remains undetermined and probably indeterminable. Of course, the specific form of the CCI and the TCI will be found to be wrong as research teaches us more about human judgment. But what I insist will *not* be wrong is the demand for specification of both task parameters and cognitive parameters and their interaction. That is the goal of the approach taken here. Achieving that goal will help us to avoid the circularity and nonfalsifiability that frequently appears in the conventional approach to understanding judgments under "stress."

Achieving that goal will also make it possible to achieve generality. For specification of task parameters makes clear exactly what conditions were and were not included in the research. For example, were nonlinear relations among indicators included? Were inverse relations included? It may be argued on theoretical or empirical grounds that neither were relevant and thus need not be included in the research. That is an acceptable argument provided the theory or empirical data are made explicit. And that can happen. In their work on "micro-worlds," Brehmer and Dörner (1993) take exactly this position. They argue that by including a set of circumstances that is representative of those to which they intend to generalize (problem-solving by public officials) their conclusions are secure; all that is necessary is to defend their practices for inclusion. That approach is what Brunswik (1956) called "canvassing," a less than perfectly rigorous form of situation sampling that makes its advantages and disadvantages clear.

Evolutionary Significance

The parallel character (inclusion of the same set of parameters) within the CCI and the TCI is noteworthy for it demonstrates my adherence to the idea that evolution would induce the development of a cognitive system that would parallel the ecological system in which hominids and Homo sapiens apparently evolved (see Hammond, 1996, for further remarks on evolutionary psychology).

These five premises, together with the inclusion of the two metatheories of truth, and the constancy theory described in chapter 6, form the foundation on which the present work is based. This foundation, however, requires that we take the properties of task conditions (particularly their range and variation) as seriously as we do cognitive conditions. That will require an explicit theory of task conditions. Whether the theory that I have provided here will be useful remains to be seen. The first test of this theory will come in its ability to enhance our understanding of task-cognition interactions, a test to which I now turn in chapter 8.

8

Predictions of Cognition-Task Interaction Under Destabilization

In this chapter I address the problem of predicting the cognitive consequences of the disruption of a person's stabilized relations with a task environment. I take into consideration the two metatheories of coherence and correspondence, three types of tasks that are encountered by human beings, and three forms of cognition they bring to bear on them. I then examine the specific ways the two major forms of disruption, endogenous and exogenous, change cognition, as well as the consequences of such cognitive changes for successful action.

These steps are taken as a direct consequence of all the ideas that have been put forward in the previous chapters. The reader should now know what is meant by "the two metatheories of coherence and correspondence" and should be able to use them to differentiate various theories of judgment and decision making. Readers should also be able to recognize "three types of tasks" as tasks located at different points on a Task Continuum Index. They should also be able to recognize "three forms of cognition" as cognitive functions located at different points on a cognitive continuum. And readers should be able to distinguish between "endogenous" and "exogenous" forms of disruption. With that background, readers will be prepared for a discussion of how such disruptions affect task-cognition interaction, and the consequences of specific forms of interaction. For that has been the goal of this book; to enhance understanding of the effects of what is commonly called "stress" on human judgment.

All this makes for a complex presentation, but it is my firm conviction that such complexity is necessary if we are to avoid oversimplification of the problem of understanding the nature of human judgment, particularly under conditions of destabilization. If we are to develop a cumulative science of human judgment, we must take into consideration the nature of different tasks, the different forms of cognition that are induced by them, as well as the different ways in which stabilized relations between the person and the task can be disrupted.

My procedure will be to provide a schematic diagram (separately for each metatheory) that will show how different cognitive activities (intuition, quasi rationality, analysis) are to be related to three different types of tasks, and what is to be expected under different types of disruption. Such a diagram can be seen in Figure 8.1.

Figure 8.1 presents a general prediction for those situations in which Cell 9 describes the normal operating situations in, say, a control room of an airplane, a ship, a power or processing plant. Here I am assuming a physical system that is under the rational, analytical control of engineering logic, that is, under the control of knowledge derived from science and technology; it is functioning in coherent form. Similarly, with regard to the operator, she or he is operating the system in accordance with training rules. Now an *unanticipated* event occurs, one for which there has been *no* training and for which therefore there are no standard responses. In plain language, the operator doesn't know what to do; disruption has occurred.

The episode of UAL 232 offers a clear example. At the start of our analysis, the flight system is operating normally (Column A), and the

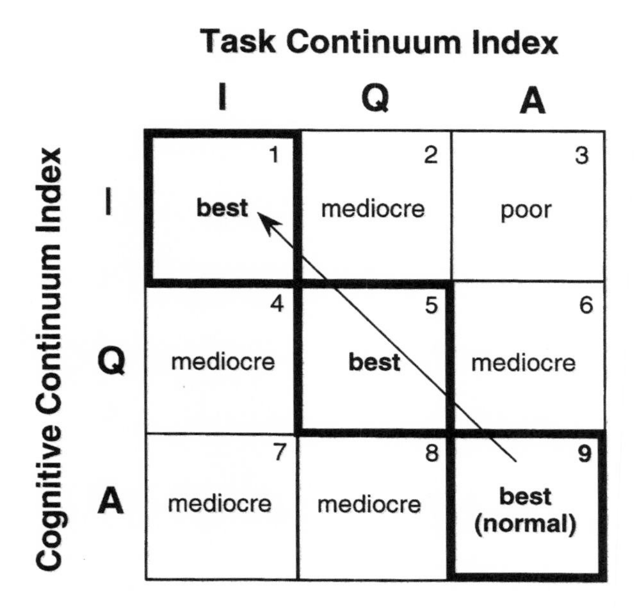

Figure 8.1. Response to endogenous disruption: from analysis to intuition.

operators of the system are operating normally (Row A). Therefore, the event begins with circumstances in Cell 9 in Figure 8.1. An unanticipated event then occurs for which there is no *prior* prearranged response. There is no emergency plan for this event. The operators' cognitive efforts are thus twice deprived; on the one hand, they are deprived of normal technological support from *task coherence* because of the technological breakdown of the system; on the other hand, they are deprived of *cognitive coherence* because sufficient knowledge to cope with the event is not available, either through training or expert assistance; help cannot be obtained from experts because no resource has anticipated this event. As a result, the operators do not know what to do, a fact that is made plain by Captain Haynes's communications with the control tower. The airplane is now out of control.

Deprived of both task coherence and cognitive coherence, the operators cannot remain in Cell 9; they must move their cognitive activity to make it congruent with the new task conditions. And these are now almost entirely intuition inducing. That is, the operators must resort to dependence on multiple fallible indicators to tell them what the aircraft is doing and where they are going; the operators have proved that it is impossible to bring analysis to bear on their circumstances. The disruption of normal Column A and Row A conditions have resulted in task conditions moving to Column I, and thus driving the operators to Row I. Task–cognition interaction has changed. Does that mean that degradation of performance is inevitable? No, for intuitive cognition is now congruent with intuition-inducing task conditions. It will be the robust properties of the weighted average linear organizing principle that will make survival possible. For it is in precisely these situations, in which only multiple fallible indicators are available, in which the weighted average organizing principle is robust (see above discussion of the linear organizing principle).

This explanation is consistent with the excellent performance of the crew of UAL 232. They gave up their analytical efforts to restore the *coherence* of the technological system and turned to orienting the aircraft to *correspond* with the empirical demands of flight. They turned to the perceptual-motor task of achieving the appropriate orientation by visual means, and thus managed to get the airplane to an airport.

But not everyone who is driven from analysis to intuition will be driven toward success. If operators are operating in Cell 9 (where analytical cognition is applied to an analysis-inducing task) but circumstances change to make it impossible to continue to function completely analytically, success or failure of the resultant cognitive activity will depend on the nature and extent of the change in the task environment.

The problem of whether to launch the *Challenger* space shuttle provides a second example of a situation in which the application of analytical cognition was frustrated, but one with a different outcome, and therefore one that deserves our attention. My interpretation of the cognitive activity employed in the launch of the *Challenger* (based upon the *Report of the Presidential Commission on the Space Shuttle Challenger Accident* [June 6, 1986] is as follows.

The process for deciding whether to launch was clearly a fully analytical one. Criteria for launching based on engineering facts and physical science had been developed over a long period of time. Every criterion had been justified by open critical review over a period of years. In the process of the launch of spacecraft, perhaps more than in any other human endeavor, each and every step had endured critical review by numerous scientists and engineers. They had constructed a clear, perfectly defensible (and defended) model of the launch procedure. It was all perfectly retraceable. Moreover, the model was to be applied to a well-understood, well-modeled environment under the control of well-known physical laws informed by many accurate instruments. Nothing was left to chance; all uncertainties, if not eliminated, were reduced to controllable limits, and if circumstances were not within these well-understood, controllable limits, there would be no launch. Only the weather remained outside these limits.

At 1:00 P.M. on January 27, however, suspicions began to arise that these limits were going to be approached. These suspicions surfaced in a telephone call from a manager to engineers at the company that constructed the booster for the shuttle. He wanted to know whether they "had any concerns about predicted low temperatures" at the launch site (Presidential Commission on the Space Shuttle Challenger Accident, 1986, p. 104). That telephone call led to a series of virtually nonstop calls and meetings among engineers, managers, and between these groups until the time of the launch the next day at 11:38 A.M. EST. The issue seemed to be straightforward: Would the "O rings" on the booster *seal,* as they must, at the low temperatures forecast for the launch? If yes, launch; if no, do not launch.

But there was a difficulty. The O-rings had not been tested directly at the predicted low temperatures, nor had previous launches been made under these low temperatures. In short, the decision makers were encountering the situation in which a fully justified analytical model could not be directly applied to a new situation because of an absence of data. As a result, the decision makers were now moved from Cell 9 (analytical environment–analytical cognition) to Cell 5 (quasi-rational environment–quasi-rational cognition). Cognition lost its fully analytical character and became quasi-rational because the carefully built model of step-by-step decision making for the launch could no longer be applied in all its justifiable detail. The data necessary for its application were not available. That is, the model of the launching procedure could no longer be *completely* applied because the environment now provided data that could not be usefully incorporated into the launch model; there was no context for the unanticipated low temperatures. As one high-level participant put it, "We said, gee, you know, we just don't know how much further we can go below the 51 or 53 degrees or whatever it was. So we were concerned with the unknown" (p. 94). Just as the pilots of UAL 232 found that they could not rely on a previously justified model of action and were thrust back on less-than-fully justified cognition, the engineers, scientists, and managers of the *Challenger* launch found that they could no longer rely on fully justified cognition for the decision making process for the launch. At that moment, analytical cognition became quasi-rational cognition.

The reason cognition became quasi-rational, that is, moved from the analytical pole but not moved all the way to the intuitive pole of the cognitive continuum (as in the case of the pilots of UAL 232), is that the use of unadulterated intuition would have meant completely unjustifiable, nonretraceable judgments of when to launch. In contrast to the pilots of UAL 232 who did not have to justify each adjustment of the throttles that were controlling their airplane, those responsible for the launch of the *Challenger* would have to provide *some* analytical justification for the order to launch—or not to launch. Their highly visible situation would not allow anyone to say, "Well, I just had this gut feeling that it would be ok to launch" (or not to launch). Thus, the decision makers found themselves forced to be as analytical as possible and as intuitive as necessary, in a word *quasi-rational.*

That circumstance was brought about because they were forced to cope with *irreducible uncertainty for which there was no analytical model available.* More specifically, no *substantive* analytical model was available because the temperature data were beyond the limits of the launch model. People whose primary function—and pride of profession—is to *drive out* uncertainty, dread irreducible uncertainty; and irreducible uncertainty beyond the limits of a model is anathema.

But some decision had to be made. It had to be "launch or don't launch." There were, however, *procedural* models available, if there were no substantive models available. That is, those responsible for the launching decision could have—and in my opinion—should have, turned to a procedural model. The moment for that occurred when the obstacles to the use of analytical cognition became stubbornly apparent. And that moment became apparent when the vice president for engineering at the manufacturing company reported that he had been asked "to take off his engineering hat and put on his management hat" (p. 94). This request came about because a committee had conducted an "unemotional, rational discussion of the engineering facts as they knew them at that time; differences of opinion as to impacts of those facts, however, had to be resolved as a *judgment call* (italics added) and therefore a management decision" (p. 93). Thus, quasi-rational cognition was introduced into the decision process and described by the participants as a "judgment call." As result, the vice president changed his position from "no launch" to "launch."

Note the use of the term "judgment call." It signifies that a *different* form of cognition would be employed, different, that is, from the "unemotional, rational discussion" that was no longer possible because of the loss of the analytical model. In short, the cognitive activity of the decision makers would have to change, would have to depart from fully analytical cognition. What would it change to? And what would control the process of change?

Recall that the vice president began thinking as an engineer (assessing the facts in a fully analytical manner that demanded coherence). During that process he discovered that the facts regarding temperature could not be used by the carefully developed analytical model. He was then requested (not required) to shift from the engineering mode to a management mode of cognition, and that, he acknowledged, would require his *judgment*. Thus, deprived of his analytical base, being forced to rely on his judgment, means that he would have to include *some* intuitive components in his cognitive activity and thus engage in quasi-rational cognitive activity. And as is always the case when intuition is employed, his cognitive process would thereby become less than fully retraceable. The vice president would no longer be able to justify fully his decision, whatever it would be, for it would have become a "judgment call." A complete retracing of his cognitive activity would now become forever impossible because there was no procedural model, no model for the quasi-rational judgment—the "judgment call" that would control the launch. As a result, there would now emerge an industry devoted to attributing various motives, good and bad, for the vice president's decision to launch.

The reader, however, will not have to descend into the question of the vice president's motives. For the reader can turn to Figure 8.1 and find a cognitive explanation, one that does not seek grounds for assigning good or evil. For although the task—deciding to launch or not to launch—demanded the fully analytical cognition of Cell 9, changes in task circumstances would not permit it. Cognition was *forced* to move to Cell 5—a combination of analysis (engineering facts insofar as they *could* be applied) and intuition (insofar as it *had* to be applied). As Figure 8.1 indicates, this was the "best" (but not perfect) procedure. Had the vice president moved to full intuition, then he would not be aware of, and thus could not justify any of his cognitive processes. Had he maintained a fully analytical mode (stayed in Cell 9) he would have had to apply a newly created, untested—and thus empirically unjustified—model or apply the current model (already found to be inadequate) incorrectly.

But, the reader may ask, if this was the "best" mode of cognition for the task, why did the disaster occur? It occurred because the quasi-rational cognition employed by the manager should have been better than it was. That much is obvious; the *Challenger* went down. *Could* it have been better than it was? Yes; there are two possibilities. If, somehow, the decision makers had made some intuitively derived judgments about conditions that were more accurate than the ones they made, they may have delayed the launch. This first possibility is another way of saying that if the managers had been more skilled in their engineering intuition, the launch would have been delayed. But, of course, there is no way of knowing whether *any* managers would have, or could have, been more skilled.

A second possibility is that if the managers had been aware of, or trained in the use of, *procedural* (decision making) models, the managers could have made better quasi-rational judgments than they did. Specifically, a procedural model developed prior to the launch date would have anticipated that events would occur that would introduce irreducible uncertainty and thus induce quasi-rational cognition—indeed, make it inevitable. If so, a less than-fully-defensible procedure would have been developed and held in readiness. That procedural model, even though not fully defensible, could have been reviewed and criticized prior to a decision to launch or delay. In all likelihood, it would have contained more analytical components than the "judgment call" made by the vice president at the time of the launch because more time would have been available for review and criticism. As a result, the managers would have found themselves defending a cognitive decision procedure, rather than stupidity or venality; everyone would have been better off.

What would that procedure look like? It would begin by acknowledging Type I and Type II errors, that is, the error of doing something when one shouldn't have (in this case, launching), and the error of not doing something (not launching) when one should have. The next step would have been one in which the probabilities, and particularly the *values,* for each decision (correct launch, correct abort, incorrect launch, incorrect abort) would be specified a priori. Had that been done as explicitly and analytically as possible, one can assume that the negative values of an incorrect launch would have swamped the decision process, and as a result, the *Challenger* would not have been launched. Be that as it may, the significant point is that the decision maker's *facts* (the probabilities of success and failure) and *values* would have been explicitly separated, made apparent for all to see. (For textbook prescriptions of the specifics of such models, see von Winterfeldt & Edwards, 1986; Yates, 1990; see Hammond & Adelman, 1976, for an application; see also Cooksey, 1996; Hammond, 1996, for further examples of quasi-rational judgment analyses.)

Criticism would have been vigorous no matter which decision would have been made; that is the price that must always be paid for quasi-rational judgments and decisions that control people's lives. The advantage gained is that by virtue of introducing as many analytical components as possible prior to the launch, a significantly larger part of the process could have been made visible and thus justifiable. The inevitable attacks would then be made on the cognitive elements of the process, and it would become necessary to attribute imperfect personal motives to the managers.

Thus, Figure 8.1 offers testable predictions about cognitive activity for persons whose circumstances begin with a disruption of Cell 9. In the case of UAL 232, disruption of constancy caused by *both* endogenous and exogenous factors led the operators (pilots) to move toward and eventually reside in Cell 1, and as a result, employ intuitive cognition with considerable success. The launch of the *Challenger* also began with all concerned residing in Cell 9, but disruption of constancy by, again, both endogenous and exogenous factors, deprived the decision makers of analytical competence. In that case, however, the decision makers were induced to move only as far as Cell 5, and thus to the use of quasi rationality rather than to the use of intuition. Success was denied them, however, because the quasi-rational process was not as competent as it needed to be, and that was due to the lack of a defensible quasi-rational decision making model prepared in advance that would have removed the necessity for the nonretraceable judgment call.

Starting Points Make a Difference

Not all decision makers are located in Cell 9 to begin with, however. Figure 8.2 offers predictions about the cognitive activity of persons who begin in Cell 1. These are the circumstances in which persons are residing in an environment that is initially intuition-inducing (Column I) and whose cognitions have thus been induced to be intuitive (Row I). The primary difference between the predictions in the two figures is that persons in Cell 1 in Figure 8.2 may be driven in different directions (across the rows or down the columns), depending on context. Consider first a decision maker who is offered assistance by a decision analyst. Suppose, for example, that a chief executive officer (CEO) has been making judgments and decisions largely intuitively and in relation to highly uncertain circumstances. In short, suppose that his starting point is in Cell 1.

Endogenous Disruption

Suppose now that there is an endogenous disruption; say, the CEO loses a number of staff members on whom he has depended to provide him with information, or their judgments, about certain sectors of his firm's activity. (This is analogous to a pilot suddenly finding that a number of instruments are inoperative.) Turmoil ensues because the CEO assumes that his judgments will be less accurate (correspondence constancy will be lost) and its restoration is demanded immediately. Now assume that

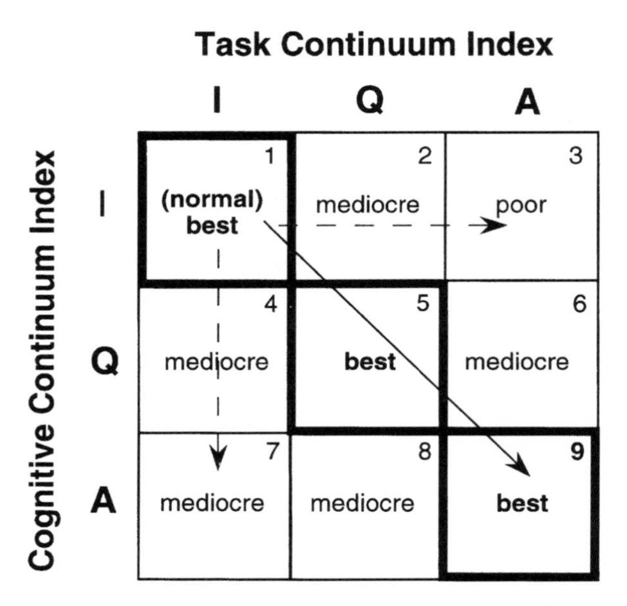

Figure 8.2. Differential responses to endogenous disruption: from intuition to?

a cry for help—the CEO will claim "stress"—is answered by a decision analyst, not an unlikely assumption today. What will the decision analyst do? The person will first try to "structure the task" as much as possible, thus moving the task environment as far toward Column A as possible. This move will be unproductive, however, unless the decision analyst also moves the CEO as far toward Row A as possible, that is, he encourages the CEO to be as analytical as possible. Since the decision analyst will not be able to (1) reduce all uncertainty in the environment, nor (2) eliminate all intuitive aspects of the CEO's judgments, the decision task is now located in Column Q. Therefore, the decision making process will move from Cell 1 to Cell 5, which, as indicated in the figure, may, in fact, be one of the best places to be. Success will not be guaranteed, of course. However, a *partly* retraceable, defensible, quasi-rational process will have been achieved, and that is the best that can be achieved in conditions that remain irreducibly uncertain to some degree. The CEO will have become as analytical as possible, and as intuitive as necessary.

As Figure 8.2 indicates, however, different people in the same circumstances may be driven in different directions from Cell 1 under endogenous disruption, as the case of the smoke jumpers (described in chapter 6) demonstrates. At the time, the foreman and his crew were working their way down Mann Gulch toward the fire, they were engaged in intuitive cognition (i.e., responding to a natural environment that was intuition-inducing, that offered multiple fallible cues—i.e., grass, trees, rocks, and terrain slope). And when the unanticipated disrupting event occurred—the fire began to rush *at* them—they continued to be intuitive cognizers; they did "what comes naturally"; they ran away from the fire, the foreman included. They stayed in Cell 1. Soon, however, the foreman sees that the fire is going to overtake them. He stopped and, in a manner inexplicable to us, shifted directly to Row A—*creative* analytical cognitive activity. That is, he rapidly analyzed the physical features of his environment—the forces generated by the wind and flames—looked at the terrain immediately underfoot, and responded in an if-then manner, characteristic of analytical cognition. That is, *if* I light a fire immediately in front of me, *then* I can lie down in its ashes and escape the flames. Note that for him, the task environment had also shifted to Column A; he was no longer in the grip of multiple fallible indicators; he was *analyzing* the circumstances in terms of time and distance and heat.

This will be a hard conclusion for almost everyone, lay persons as well as psychologists, to accept; the premise that all creativity is a product of a mysterious cognitive activity, usually described (but never defined) as intuition, is so commonplace that it is seldom if ever questioned. Thus, to assert that someone was analytically creative under "stress" is to defy traditional belief. But that is precisely the role of theory, that is its function—to drive us to counterintuitive conclusions that challenge "what everybody knows," even if the conclusions remain only as doubtful hypotheses. And that is also the role of well-documented reports of behavior under stress (such as the foreman's behavior at Mann Gulch). They should drive us to reexamine our well-accepted (but not well-tested) beliefs, as, for example, that stress "narrows the cognitive field," and thus to expand the range of our theoretical conjectures. Naturally, I hope that the above paragraphs have encouraged the reader to expand the range of his or her theoretical conjectures, an activity badly needed in the study of judgment under stress. The above conclusion becomes more compelling when the foreman's behavior is contrasted with the behavior of the rest of the crew.

It should be recognized that the smoke jumpers were not naive with respect to this type of threatening event; they had received specific training regarding abrupt explosions of this sort. Their instructions were to (1) backfire if they had time to do so and the situation was right for it (neither condition applied in Mann Gulch); (2) turn back into the fire and try to find havens within it (impossible in Mann Gulch because of the wall of flame); (3) not let the fire determine where it would get at them (the terrain prevented this choice); and (4) get to the top of a ridge where there is less fuel for the fire to burn (which they tried to do); it was the only option that did not require "detour behavior"; it offered direct escape. All the smoke jumpers took the fourth option, but it was a useful option for only two of the 15. The following description of the fire scene by a Forest Service official the next morning makes the reasons clear:

> The ground appearance was that a terrific draft of superheated air of tremendous velocity had swept up the hill exploding all inflammable material, causing a wall of flame (which I had observed from below at 5:30 p.m. the previous evening) six hundred feet high to roll over the ridge and down the other side and continue over ridges and down gulches until the fuels were so light that the wall could not maintain heat enough to continue. This wall covered three thousand acres in ten minutes or less. Anything caught in the direct path of the heat blast perished. (Maclean, 1992, p. 120)

Creative Analytical Cognition

Under these circumstances it is hard to see how anyone would do anything else but run from the ridge; surely, any other activity would demand analytical, "counterintuitive," action, that is, going *into* the fire, and that would be unimaginable under these circumstances. But that is exactly what the foreman did; he changed his cognitive activity from one form to another, from intuition to analysis. And he exhibited creativity in the process of doing so. For when asked at the hearing "whether he had ever been instructed in setting an escape fire," the foreman replied, "not that I know of. It just seemed the *logical* thing to do" (Maclean, 1992, p 101; italics added). Note that Maclean sees the point in my terms, for he says, "Being logical meant building one fire in front of another, lying down in its ashes, and breathing close to the ground on a slight elevation. He relied on a logic of a kind and the others [relied] on time reduced to seconds" (p. 101). By that Maclean means that the intuitive cognizers relied on their intuitive judgments of the speed of the fire up the slope, the distance to the top of the ridge, and their ability to run that distance in the few seconds the fire would allow them. Their judgments were wrong.

Dodge's description of his fire is mostly from inside it.

> After walking around to the north side of the fire I started as an avenue of escape, I heard someone comment with these words, "To hell with that, I'm getting out of here!" and for all my hollering, I could not direct anyone into the burned area. I then walked through the flames toward the head of the fire into the inside and continued to holler at everyone who went by, but all failed to heed my instructions; and within seconds after the last man had passed, the main fire hit the area I was in. (p. 99)

When asked at the hearing if any of the crew had looked his way as they went by, he said no, "They didn't seem to pay any attention. That is the part I didn't understand. They seemed to have something on their minds—all headed in one direction" (pp. 99–100). Maclean's descriptions of the foreman's judgment and decision making show an unusual grasp of that activity. For example:

> [The foreman's] running but not his thinking stopped when he saw the top of the ridge, for he immediately thought his crew could not make the top and so he immediately set his escape fire. When he tried to explain it, it was too late—no one understood him; except for himself, they passed it by. Except to him, whom it saved, his escape fire has only one kind of value—the value of a thought of a fire foreman in time of emergency judged purely as thought. The immediate answer to the storyteller's question about the escape fire is yes, it was strange and wonderful that in this moment of time when only a moment was left, Dodge's head worked. (p. 103).

Any comment on Maclean's description of the foreman's thought would surely be superfluous; few psychologists would be as insightful. But he didn't stop there; he also sees the difference between the foreman and the crew: "His invention, taking as much guts as logic, suffered the immediate fate of many other inventions—it was thought to be crazy by those who first saw it. Somebody said, 'To hell with that,' and they kept going, most of them to their deaths" (p. 106).

It will be useful to compare the smoke jumpers' situation with the one on the USS *Samuel B. Roberts*, as it is described by Commander Rinn. Here the man in charge also came up with a deviation from what was expected, in his case a deviation from doctrine. But in contrast to the foreman's crew, the navy crew does exactly what Commander Rinn orders them to do.

And that is what should have been done, according to the theory put forward here. For the theory argues that in the face of exogenous disruptions to constancy—as the fire and explosions certainly were—operators should continue to do what they are trained to do; they should not try to be inventive or creative, as for example, the smoke jumpers' foreman was (see my description of the Aloha incident in chapter 6.) It is true that the captain of the ship deviated from naval doctrine, but that simply meant that he chose a *different* tactic; he didn't invent one. Exogenous disruptions, in short, test training (Was it the right training and was it successful?) and discipline (Did the trainees do under adversity what they were trained to do?). Creativity seldom, if ever, has a place in response to exogenous disruptions.

To be fair to the smoke jumpers, their situation was perhaps more demanding of intuition, if not creativity. The roaring fire was overwhelming them with noise and tremendous heat and flames, and they were running for their lives. But had they been as well trained and disciplined as the navy crew was (Maclean finds it significant that the foreman and the crew had not trained together), they may well have given the foreman's change in cognitive activity, his logic, more credence, perhaps followed his advice, and saved their lives in this circumstance involving endogenous disruption.

Research Linking Destabilization to Cognitive Change

The following experiments were planned and conducted without my knowledge; I learned of their existence only when I read the published articles. In that sense they are truly independent tests of the theory.

Mahan (1994) has conducted several laboratory experiments within the context of the theory described here, albeit prior to its explication in its present form. He makes only a rough distinction among tasks with regard to their intuition and analysis-inducing properties, and similarly with regard to the cognitive activity of his subjects. Nevertheless, his work is valuable because it advances our understanding of Cognitive Continuum Theory in relation to stressful circumstances.

Mahan began with the observation that continuous work is more and more frequently demanded of persons who control processing plants, utilities, etc., and because continuous work leads to fatigue, which would have a destabilizing effect, he therefore decided to inquire into the effects prolonged work might have on judgment and decision making. And because he chose Cognitive Continuum Theory as a theoretical framework, he studied the effect of continuous work (4 hours) in relation to (1) an intuition-inducing task (integrating information from multiple fallible indicators, a task that simulated production forecasting), (2) analysis-inducing tasks (mental arithmetic, syllogistic reasoning), and (3) the degrees of irreducible uncertainty were varied in the latter tasks. Mahan argued that fatigue-producing continuous work would have differential effects on cognitive activity in these two types of tasks. He reasoned as follows:

> Given the aforementioned considerations, it is reasonable to perceive that task uncertainty and continuous work may be stressful conditions under which to perform a task. The effects of this stress are likely to be manifested as a modification of the strategies used by subjects to complete a given task. However, changes in performance strategies should be task specific. Taking the cognitive continuum perspective as a framework in which to predict the effects of uncertainty and continuous work on the judgment process, one could anticipate a shift toward an intuitive mode of processing under uncertain task conditions and during extended performance. This shift would reduce the intellectual effort necessary for task performance. A behavioral index of intuitive cognition that is based on a relatively rapid form of processing is that, in theory, one should observe more judgments per unit time than in an analytically based mode of cognition. Further, task uncertainty and the stress of continuous work should tend to interact. Thus, the shift toward an intuitive mode of cognition could be moderated by the uncertainty of the task. In this case, the shift toward intuitive cognition should be more apparent under greater levels of task uncertainty through the observation of more judgments per unit time with a corresponding reduction in the accuracy of those judgments, the latter of which reflects the application of approximating strategies in performance. (Mahan, 1994, p. 94)

But different behavior should be observed in relation to tasks demanding analytical work.

> In distinct contrast, the strategy shift in response to the stress of continuous work in analytical task performance should be seen as a reduction in the rate of problem solving. Because the analytical tasks are well defined and the subjects have, in theory, control over the organizing principle for task performance (i.e., the axioms of simple addition and the logic of simple syllogistic reasoning), precision in performance should be maintained but at the expense of number of problems attempted. Thus, these qualities of the task should produce performance rate decrements over the duration of the 4-hr. continuous work study. (p. 95)

Mahan's results support these hypotheses. More judgments per unit time were made in the intuition-inducing task as work continued, and in the analysis-inducing task the rate of problem solving decreased over time.

Mahan also inquired into the effects of various degrees of task uncertainty in information processing. Although previous researchers such as Brehmer (1974), Brehmer and Lindberg (1970), and others (see especially Brehmer & Joyce, 1988) had found that task uncertainty was directly related to difficulty in learning, as Mahan points out, "it remains less clear how task attributes such as uncertainty and continuous work alter the cognitive strategy manifested by an individual during task performance" (1994, p. 94). Mahan's results are clear: In the intuition-inducing task involving multiple fallible indicators, increasing uncertainty decreased the *accuracy* of the subjects' performance, decreased the subjects' *consistency* of judgment, and *increased* the rate at which judgments were made. That is, "as task duration increased, mean values on achievement and consistency measures became smaller, whereas mean values on judgment rate became larger" (p. 104). Further, it is clear that these changes in the subjects' *knowledge* of the task characteristics did not cause these changes. Thus, task uncertainty and task duration affect *execution* or *performance,* not knowledge (thus replicating Rothstein, 1986). Also, "subjective reports on perceived stress, sleepiness, and negative mood . . . suggest that continuous work and task uncertainty were indeed viewed by subjects as stressful conditions under which to perform" (Mahan, 1994, p. 106).

The second part of Mahan's hypothesis concerned continuous work in analysis-inducing tasks (mental arithmetic and syllogistic logic). As anticipated, he found that "subjects were able to manifest a high degree of accuracy . . . for the entire study. However, it was . . . evident that the *rate of performance* [italics added] on these tasks deteriorated over time" (p. 108) (see also Mahan, 1991, 1992).

> Presumably, subjects reduced their rate of performance in order to maintain their accuracy. Given the deterministic quality of the analytical tasks, perseverating over these problems (i.e., gathering one's thoughts, etc.) may have helped subjects to generate higher precision. It is possible that

> the behavioral effect of fatigue may be found here in a rate-reduction strategy that allows subjects to exert better control over the application of the axioms of arithmetic and logical reasoning. Reductions in problem solving rate on subject-paced tasks is a typical response to fatigue and reduced arousal (see Broadbent, 1971). (Mahan, 1994, p. 108)

To sum up: In Mahan's study continuous work on intuition-inducing tasks resulted in *decreasing quality* of performance and *increasing rate* of performance, both exacerbated by higher degrees of uncertainty. Continuous work in analysis-inducing tasks, on the other hand, did not reduce the quality of performance but did decrease the rate of performance. Analysis takes time, and, as Mahan points out, subjects apparently guarded the quality of their performance by slowing down.

Mahan's work is important because it (1) illustrates the testability of the theory, (2) confirms the utility of the basic parameters of the theory for *predicting* different behaviors in different tasks, and (3) does so in the context of an important environmental condition—continuous work.

Summary

This chapter represents the culmination of an effort to develop a coherent theory of judgment under what is commonly called "stress." In addition, it offers some examples of how the putatively coherent theory can be put to empirical test, and thus achieve correspondence with the facts of behavior. In order to arrive at this culmination, I have tried to build directly on the work in previous chapters in part II. That is, I have tried to be specific about why the term "stress" is not a useful one for our topic, but that "disruption of constancy" is, why we must explicitly consider the metatheories of coherence and correspondence, why we must differentiate among judgment tasks according to a theory of tasks, as well as why we must examine different forms of cognition according to a theory, and why we must consider different types of disruption without regard to their content (such as fire, noise, etc.). As a result, I hope that this chapter will encourage researchers to pursue the new directions that it offers.

9

Moral Judgments Under Stress

Can moral judgments be studied within a scientific framework? That question creates a serious problem for students of behavior; it directly challenges the limits of behavioral science. This challenge has frequently been made and can be clearly seen in the following remarks by Earle (in press) who places the study of moral judgments *outside* the purview of science:

> Moral judgment and behavior are uniquely resistant to psychological analysis because morality is traditionally construed in terms which do not admit of control by psychological laws. In particular, no action can be termed moral without thereby invoking the idea of *freedom*. An agent can act morally only on the condition that it is free. . . . In the fully emancipated state envisioned by the philosophers of the Enlightenment, the modern, autonomous individual need appeal no further than to the court of its own rationality for justification of its . . . conduct in the world around it. Thus, regardless of whatever the psychological dimension of moral behavior may in fact be, there is no way around the purely conceptual dilemma that to account for moral behavior psychologically is to deny such behavior the exacted autonomy upon which its moral character is predicated; with the result that it is *ipso facto* no longer properly *moral* [italics added] behavior under consideration.

I will return to this issue at the end of this chapter. First, however, we need to confront other issues.

In short, when we move from the matter of the effects of stress on professional-technical judgments to moral judgments we move from muddy waters to a morass. For whereas analytical cognition is generally assumed to be the most appropriate mode of cognition in the realm of professional-technical judgments—no one wants the persons running the nuclear plant to let their intuitions tell them which valve to turn—moral judgments invite, encourage, and may even demand, intuition. The basis of professional-technical judgments may not always be perfectly clear to us—they, too, may include intuitive components and thus become elusive, hard to follow, and difficult to understand—but they have the advantage of generally appearing in a hard, empirical context. The judgments of airline pilots, naval officers, nuclear plant operators, and firefighters can be evaluated in terms of concrete events and the behavior demanded of the situation. We may be confused—details may be lacking or contradictory—but we usually have a fairly clear idea of what was intended, what was expected, and thus can pinpoint what went wrong. We are less certain when the judgment contains moral components. Were the rescuers impartial? Were the resources awarded to those who needed them most? Readers will be all too familiar with the moral and ethical issues that arise from professional-technical judgments made (post hoc) on stressful occasions.

Consider this situation:

> Three members of an expedition of Indo-Tibetan Border Police, who were climbing from the Tibetan side of the mountain [Everest], were discovered half-frozen by a Japanese team who, in their lust to reach the summit, simply passed them by. One of the Japanese climbers blithely told a British journalist, "We didn't know them. No, we didn't give them any water. We didn't talk to them. They had severe high-altitude sickness. They looked as if they were dangerous." Another said, "Above 8000 meters is not a place where people can afford morality." (Fraser, 1997, p. 63)

Thus, we see that *professional-technical* judgments can often stand up to emotions induced by external circumstances that threaten life, whereas *moral* judgments can deteriorate in those same circumstances.

No one was watching when those Japanese climbers passed by the other climbers, but even in the most public places moral judgments create problems. Patrick Henry, famous for his cry of "give me liberty or give me death," provides a historical example of how a great champion of liberty addressed the question of whether the new American Constitution should outlaw slavery. After declaring his noncompensatory belief in "liberty," he realized he had to choose between accepting slavery or creating chaos within the Union. Speaking in Virginia *against* ratification of the proposed Constitution Henry declared:

> We deplore [slavery] with all the pity of humanity, [but] as much as I deplore slavery, I see that prudence forbids its abolition. . . . It would rejoice my very soul, that everyone of my fellow beings was emancipated. . . . But is it practicable by any human means, to liberate them, without

producing the most dreadful and ruinous consequences? Their manumission is incompatible with the felicity of the country. (O'Brien, 1996, p. 72)

Thus, Henry framed his demand for "liberty" as a value that was noncompensatory for *him*. Nothing could compensate him for its loss, hence his declaration of "liberty or death." But it was another matter altogether when the liberty of the slaves was at stake. On the one hand, he strongly wished for the liberty of the slaves (we should take him at his word) but, on the other hand, when he judged that their liberty was "incompatible with the felicity of our country," he chose the "felicity of our country" over liberty for the slaves. It is easy to sneer at Henry for his (possible) hypocrisy. But haven't we all been forced to choose between incompatible values—justice versus forgiveness, for example—and suffered the stress and embarrassment of being reminded that we have abandoned one in order to achieve the other?

It also requires "moral courage" to be as frank as Patrick Henry was; it is not often that we openly announce that we are going to relinquish one common and deeply held value in order to achieve or preserve another one. But very often "moral courage" means doing just that. It takes moral courage to cope with the disruptions caused by the appearance of a competing value that cannot coexist with the one already accepted.

In short, moral judgments are of a different order from professional judgments. It is generally much harder to reconstruct problematic moral situations (observations and recollections of them will differ), the ensuing behavior is much harder to describe and recall (interpretations of them will differ), and it will be much harder to find a consensual framework to evaluate the issues (different values will be at stake), and as a result, judgments of right and wrong will differ. In short, there are formidable barriers to understanding and evaluating moral judgments, even by the people who made them.

Can the conceptual framework that was applied in previous chapters to technical-professional judgments be extended to the understanding of moral judgments? Can it be extended to understanding moral judgments under "stress"? That is the challenge to be addressed by this chapter.

Before attempting this task, let me remind readers of the basic concepts of the theory that I will attempt to apply to moral judgments. They include defining "stress" as a loss of, or disruption of, constancy; differentiating cognitive activities in terms of the correspondence metatheory and coherence metatheory (both of which include a cognitive continuum of analysis, quasi rationality, and intuition); differentiating tasks that lie on a continuum from those that are analysis-inducing, quasi rationality-inducing, and intuition-inducing; differentiating constancy disruption into endogenous and exogenous types. (See, Figure 8.2 for an example of how these concepts are linked together.)

In order to extend the theory from technical-professional judgments to moral judgments, my intention is to employ each of these concepts in a manner consistent with their prior use in chapters 6, 7, and 8.

Defining Stress

Defining stress as *loss of constancy* under physical circumstances is a plausible definition; the empirical *fact* of constancy is unquestionable, indeed, has been unquestionable for almost a century. Nor is it questionable that maintaining constancy is a critical feature of life for a wide variety of species; survival without it seems impossible for mammals. Therefore, loss of such an essential feature of our existence certainly would cause distress. For constancy was discovered under *physical* circumstances, the physical size of this object, the physical shape of that one, the light rays of color from another one. Can we arrive at a plausible understanding of what "constancy" could mean in nonphysical, that is, social, circumstances? What could it mean in circumstances in which *moral* judgments of right and wrong are made? Is there such a thing as constancy in moral judgments? These are some of the questions we must address in this chapter.

Constancy, it will be remembered, is the ability to correctly infer properties (e.g., size, color) of a distal object (e.g., a person, a house), despite variations in the proximal stimuli (e.g., light rays on the retina). There are two reasons for taking this phenomenon seriously. First, as McEwen (1998) pointed out, this ability is "critical to survival," just as Brunswik had pointed out in 1956 that the "constancies are the very essence of life" (Brunswik, 1956, p. 23). Thus, the capacity to maintain a constant perception of an object, despite variations in the information we receive about it, is a grand achievement of evolution because of the enormous power it gives us. But there is a second, seldom-mentioned reason that constancy is a grand achievement and that is because it illustrates a counterintuitive phenomenon, namely, the establishment of (almost) *perfect* relations between widely separated events—the cognitive activity of an organism and physical properties (size, shape, color) of environmental objects—despite *imperfect* relations between the mediating events and the object, and mediating events within the organism.

This unusual phenomenon is accurately portrayed by the lens model. Note that the model shows that the relation between the judgment and the property of the object judged is high, despite the imperfect relation between each of the indicators of the object-property and the imperfect relation between each of the indicators and the judgment. Although this central feature of psychology has seldom been given the attention it deserves, Brunswik (1956) developed its theoretical and empirical significance for physical perception, and extended it to social perception by studying judgments of intelligence, sociability, and similar social variables

(Funder & Sneed, 1993; Gillis, Bernieri, & Wooten, 1993; see also Hammond, 1996; Cooksey, 1996, for a variety of other applications). The lens model is built on the empirical phenomena of constancy—accurate perception—achieved by means of imperfect information.

In 1955 I extended this principle beyond physical or social perception to include judgment in general and showed that it could be applied to the judgments about patients made by clinical psychologists and medical doctors (Hammond, 1955). From that time forward, hundreds of studies have shown the applicability of the theory of the lens model to the analysis of human judgment over a wide range of judgments and task circumstances (Cooksey, 1996; Hammond, 1996). It has not been applied, however, to moral judgments, in particular, moral judgments under stress.

Extension of the Definition of Stress to Moral Judgments

If the theory presented earlier is to be successfully extended to moral judgments, the fact of constancy must be found to be applicable there as well, for the theory asserts that it is the *loss* of constancy that demands our attention. Therefore, if constancy is not found, then the extension of the theory must fail.

Our common experience indicates that we generally maintain a constant moral judgment—the acceptability or nonacceptability of certain behaviors—despite changes in immediate, or close to immediate, information. For example, if we see a person beating a child, we condemn that action regardless of whether the person is a man, woman, or even another child. And we condemn that action over a wide variety of circumstances, at home, on the street, in a bus, etc. Thus, we see moral constancy over both people and circumstances in this example. But as we all know, there will be circumstances in which we *condone* the killing of children, irrespective of the gender of the killer or the specific site in which it takes place; war, of course, offers a plethora of examples. Every country has both *condemned* (in law) and *condoned* (in war) harm to children from the beginning of history. In short, the constancy of our moral judgments is far less general than the nearly universal constancy we find in our physical judgments.

There is a second difference between physical judgments and moral judgments. It is difficult to create a loss of constancy in physical circumstances; it requires extraordinary events, for example, natural disasters like floods and earthquakes, to bring it about, or an expert to bring it about in a laboratory, because our biological heritage is so strong. But it is easy to create a loss of moral constancy due to changes or ambiguities in the social environment; that can be done verbally, as I did by introducing the idea that war has justified the killing of children. The psychologist Milgram became famous in the 1960s by demonstrating that it was easy to induce a wide variety of men and women to obey orders to severely punish innocent people, even when they could be heard screaming in pain (Milgram, 1974; for a modern review, see Blass, 1991).

Thus, it appears that a disruption of physical constancy will seldom occur, but will have great consequences—at least for most people—whereas disruption of moral constancy will occur often, although the consequences may be less. I anticipate no dispute about the infrequency of the loss of physical constancy by most people most of the time. But my assertion that moral constancy is lost more frequently may be doubted. In order to cope with that doubt it will be necessary to turn to the question of how moral judgments fit within the coherence and correspondence metatheories, for everything will depend on whether the moral judgments fall within the coherence metatheory or the correspondence metatheory, or both. This is as it should be, for we saw the importance of making this distinction in connection with technical-professional judgments.

The Correspondence and Coherence of Moral Judgments

Moral judgments have a dual nature. Most philosophical theories of moral judgments are based on, and studied within, the *coherence metatheory* of cognition, yet others are related to the *correspondence metatheory* (see chapter 3). The coherence metatheory is employed when the goal is not only to describe but also to *justify* a set of principles of morality or rules of behavior, and all such justifications must be logically coherent. A moral code that contains inconsistencies or contradictions will be rejected—sooner or later—if for no other reason than that it is impossible to follow contradictory sets of principles (see Smith, 1997, for a discussion of the place of analytical coherence of thought in constitutional law.) I cannot offer a critique of the numerous moral systems that have gained the attention of scholars—the literature is enormous—but I can say this; it is apparent that none has escaped unscathed from charges of failures of consistency; none has significantly outdistanced its competitors. James Q. Wilson, an influential, well-respected modern writer on matters of social policy and the role of moral judgments makes this point in graphic terms:

> Believing, as they do, in systems of thought rather than habits of life, it is intellectuals who must find a foundation for any system by which they hope to live. The history of mankind is littered with the rubble of these systems, mute testimony to the enlightened ignorance that governed their construction: nihilism, Marxism, hedonism, artistic self-indulgence. Systems collapse faster, and with greater collateral damage, than habits; habits have lasted for centuries, surviving even the most oppressive regimes erected on the most "enlightened" principles. (Wilson, 1993, p. 221)

Thus, Wilson not only mocks the efforts of those who believe "in systems of thought," but he also compares them unfavorably with those who rely on "habits of life." He draws the distinction well between those who demand a coherent justification for a moral judgment and those who are satisfied with judgments that fulfill a "moral sense" of right and wrong.

It is the latter who form their judgments within the correspondence metatheory. They will demand that our moral judgments conform to a "moral sense" of what is "right" and what is "wrong," altogether a different criterion than logical consistency for the acceptability of moral judgments. The advocates of a moral sense do not demand coherence, but something quite different, specifically, a judgment that corresponds to those of their peers. A coherent justification is not as important as justification provided by agreement with one's peers, often vividly provided by a mob scene.

Wilson thus provides an interesting contrast between the failure of all those "systems of thought" (that demand coherence) to create a standard of morality, but he doesn't give us much detail about exactly why they "collapse faster, and with greater collateral damage, than habits." He does, however, point out that even "after the Enlightenment had begun, most men, intellectual as well as practical, lived lives of ordinary virtue without giving much thought to why they did" (p. 221). That should give us pause. For if they "didn't give much *thought* to why they were living . . . lives of ordinary virtue," then they weren't demanding coherence in their moral judgments and those of others. But if they weren't thinking about the coherence of their moral judgments, what *were* they doing when they made their moral judgments? Wilson's answer: They were doing what came naturally, *without thinking*. And what could that be? Wilson relies on Voltaire for his explanation: "Even Voltaire," he said, "believed that man's nature was fundamentally moral." According to Voltaire, "man had 'certain inalienable feelings' that constitute 'the eternal bonds and first laws of human society' " (Wilson, 1993, p. 221). Now there we have it! There *are* standards, after all. But where are they? They reside in those "inalienable feelings," in which we find "the eternal (no less!) bonds and first laws of human society." Thus, we need not be at a loss for "foundations" for our moral judgments; we can rely on our "inalienable feelings" (our subjective judgments of right and wrong), for therein lie the "first laws of human society." Our search for a standard for moral judgments is apparently over.

However, Voltaire's brave words did not end our search or provide that objective standard we need so badly in order to check the correspondence of a judgment with a criterion of morality. As the reader knows only too well, and Wilson certainly should know, the behavioral expressions of those "inalienable feelings" vary widely over time and place, and thus their morality becomes subjective, for their morality lies in the eye of the beholder. Indeed, it was precisely for that reason that the intellectuals sought standards of coherence in "foundations" that could be analytically *justified*—something clearly lacking in those "inalienable feelings."

However, we can't place *all* our faith in Wilson's assertion that premodern people "didn't give much thought to why they were living . . . lives of ordinary virtue" because, after all, he doesn't really know. Or at least he doesn't really know about early Homo sapiens. Two who do claim to know about Pleistocene people are J. Tooby and L. Cosmides (1992). And, after a good deal of careful research, including many laboratory experiments, it is their conclusion that even these early people were able to develop "social contracts" of the "if you do this I'll do that" character. Indeed, they go so far as to attribute the ability to carry out approximations to such logical processes as Modus ponens ("if p then q") and Modus tollens ("if not p then not q") to people of prehistory. In addition, Tooby and Cosmides claim that early Homo sapiens was expert at detecting moral failures, that is, cheating on the contract. They go further to claim this ability to be a product of evolution; that is, evolution selected those individuals who were the best at such detection, and as a result, detection of cheating is a skill provided by our heritage. In short, there is a plausible body of evidence that suggests that the coherence metatheory was within reach of early Homo sapiens and that it was employed in the pursuit of moral judgments. So Wilson may be wrong in his guess that people "didn't give much thought to why they were living . . . lives of ordinary virtue." Tooby and Cosmides are convinced that they *could* have.

Once we see that our moral judgments can fall within either the coherence or correspondence metatheory, our next step is to see that coherence makes demands of *internal* consistency that are extremely difficult for moral judgments to meet, and that correspondence makes demands of *external* consistency that are also extremely difficult for moral judgments to meet. Thus, it follows that moral judgments will infrequently meet the conditions of constancy demanded of them, although physical judgments frequently do. That said, we need to look at these differences between internal and external consistency more closely. And that will move us to take a closer look at the role of intuition and analysis within the coherence and correspondence points of view of moral judgments.

From Correspondence and Coherence to Intuition and Analysis

I used Voltaire's remarks to represent the "naturalistic" view of the moral sense because Wilson chose Voltaire to be his spokesman, but, in fact, the strong proponents of a moral sense were Scottish and British philosophers of the eighteenth century, particularly Francis Hutcheson and David Hume. They emphasized that our innate beliefs—our *intuitions* about right and wrong—provide an entirely sufficient guide for our moral judgments about justice, fairness, and the like. Indeed, nearly all those who urge us to depend on our "inalienable feelings" believe that it is precisely such feelings that separate us from other animals and make us human. However desirable that would be, it turns out that our intuitions differ so much among us that they do not provide a sufficient guide for our moral judgments. Not only do they fail to provide a sufficient guide for our moral judgments, but those differences in intuitions about right and wrong have led and still lead to serious trouble (murder—even mass murder—of fellow human beings), today as well as throughout history. Just as the rationalists dispute one another's claim for the impeccable analytical justification of their foundational systems, and disagree about where the truth lies, and leave history "littered with the rubble of [their] systems," so do the intuitionists disagree about how their intuitions should apply to specific acts of others, and leave a "rubble"—often bodies on a battlefield—of their own.

But there is a second disturbing feature of the concept of a "moral sense" that makes it doubtful as a basis for moral judgments, and that is its *mysterious* quality; it is difficult to know exactly what intuition means, a common weakness in all efforts to explain it. The great advantage of the rationalists' treatises on morality was that they had a definite criterion to meet—analytical justification of coherence. Although proof through hard analytical work did not guarantee acceptance, if the theory failed to meet that criterion, all else failed. But how shall we build our confidence in Voltaire's (or Wilson's, or Hutcheson's, or Hume's) assertions about "inalienable feelings"? If we were consistently unanimous in our judgments, that would help, but we seldom are. And we are often inept in our explanation of just exactly what an "inalienable feeling" is, and how it differs from other feelings, for example. The same is true for the concept of "common sense"; people are perfectly prepared to condemn others for their failure to apply it, write books about its value, and why it should be used in place of analytical cognition, all the while remaining utterly indifferent to the need for explaining just what they mean, and how they will know common sense when they see it (see P. Howard, 1994, for a strikingly gratuitous—he never defines it—modern celebration of the value concept of common sense that achieved great popularity among highly placed politicians.)

Thus, the conceptual framework that enabled us to differentiate among technical-professional judgments under stress, that drew the distinction between the coherence metatheory and the correspondence metatheory and their analytical and intuitive components, may indeed carry potential for describing moral and political judgments under stress. Why? Because that conceptual framework allows us to separate different kinds of cognitive activity; one form that demands coherence and the analytical work to establish and justify it, and another form that demands only correspondence with some criterion, relies only on intuition, and accepts the justification offered by peers rather than logic. Therefore, we should now be able to pursue the question of the effects of different forms of disruptions of constancy on each of these types of moral judgments.

First, however, I want to add to the credibility of this argument by showing that it can be applied to actual events that matter and help us to understand them. The context I chose for this task is the cognitive activity that was demanded of those attempting to create a government with a "moral sense," perhaps a unique case in the history of humankind. I undertake this description to illustrate the utility of the extension of this conceptual framework to actual events.

Rivalry Between Intuition and Analysis in the Construction of a Government

Rarely has a group of people had the opportunity to construct a new government in a new country on the basis of explicit moral principles rather than naked power. However, one such opportunity occurred not so long ago, nor so far away. In the late eighteenth century on the east coast of North America, a handful of men and women were faced with the successful outcome of their struggle to be free of the rule of England. In 1776 they had declared the right to establish their own form of government and took pains to justify that declaration in a calm and rational manner in a document that is a living force today. And it is precisely because they did boldly put forth their moral judgments that moved them to declare independence that it became a living force. By beginning their document with the statement that "we hold these truths to be self-evident," they made explicit their grounds for their historic actions.

As a cognitive scientist, I can ask: What sort of cognitive activity led to those "self-evident" declarations of "truth"? Could it be described as strong analytical cognition, the kind Wilson described as "rubble-littering history"? Or could it be better described in terms of Voltaire's concept of an intuitive moral sense that guided the effort to create the first nation explicitly based on moral principles? Seeking answers to such questions is my purpose.

THOMAS JEFFERSON AS COGNITIVE THEORIST

It was easy to find an answer in the writing of Thomas Jefferson, the primary architect of the Declaration of Independence. It is fascinating to discover that the tension between analysis and intuition did in fact play a prominent part in Thomas Jefferson's thinking. While examining his political thinking I discovered Jefferson had a cognitive theory of his own, one that is not very different from modern cognitive theory. The well-known political scientist, Garry Wills (1978), had much to say about Jefferson's cognitive theory in his book on Jefferson and the Declaration of Independence. He begins by describing a curious event that led Jefferson to explicate his theory of cognition.

While Jefferson was the U.S. Ambassador to France in 1789 he wrote what is now a well-known letter to Maria Cosway, a married lady who became his close—no one knows how close—friend in Paris. The letter was written on the occasion in which he explained to Cosway his decision not to pursue their friendship any further, and defends this decision in terms of a struggle between "Head vs. Heart." The reason the letter became well known to historians is that in it Jefferson attributes various cognitive properties to "head and heart" (largely in terms of reason and sentiment). But he does this in a surprising way that is both parallel to, yet different from, current theory and research in judgment and decision making.

The letter explains the strong place of the "Heart" relative to the "Head" in Jefferson's thinking about moral and political judgments. For Jefferson makes it clear that he believes that the *heart* is the ultimate seat of reason. That is strange, of course, because although the modern view accepts the distinction between analysis and intuition, his view that the "heart" (i.e., intuition) is the seat of reason is directly opposite to current thinking about these two predominant aspects of judgment. Wills recognizes this and takes pains to explain Jefferson's position: "The Head cannot speak of duty or virtue. . . . It is not a principle of action, but only of reflection. It weighs the means; it cannot determine the end—to use the distinction of Hume [Treatise], namely, 'Reason is, and ought only to be, the slave of the passions; and can never pretend to any other office than to serve and obey them' " (Wills, 1978, p. 278). (If this were not strange enough, as I will show below, we shall find this view urged on us by a famous Supreme Court Justice in the twentieth century.) In the letter to Cosway, Jefferson asserts that "morals were too essential to the happiness of man to be risked on the incertain [sic] combinations of the head. She [i.e., Nature] laid their foundation therefore in sentiment, not in science" (Jefferson, 1984, p. 874). And, indeed, Wills notes that Jefferson goes so far as to "attribute(s) the American Revolution itself to the moral guidance of the Heart":

> If our country, when pressed with wrongs at the point of the bayonet, had been governed by its heads instead of its hearts, where would we have been now? Hanging on a gallows as high as Haman's [a biblical reference]. You begin to calculate and to compare wealth and numbers; we threw up a few pulsations of our warmest blood; we supplied enthusiasm against wealth and numbers; we put our existence to the hazard, when the hazard seemed against us, and we saved our country; justifying at the same time the ways of Providence, whose precept is to do always what is right, and to leave the issue to him. (Wills, 1978, pp. 281–282)

Thus, Jefferson proudly attributes the success of the revolution to intuitive cognition (the "Heart") rather than to analytical calculations carried out by the "Head."

Hence, we learn what Jefferson's cognitive theory actually was, for it can be seen in his assignment of cognitive properties to analysis and intuition. Despite the archaic language, it is easy to recognize his differentiation of function for it matches similar differentiations today (see, e.g., Epstein, 1994; Hammond, 1996; Hammond et al., 1987).

And like all cognitive theorists, Jefferson extends his conceptual framework to other matters and thus makes it easier to grasp his theory. For example, he assigned the properties of "Head vs. Heart" differentially to Northerners and Southerners in this fashion (Wills, 1978, p. 283):

In the North they are	**In the South they are**
cool	fiery
sober	voluptuary
laborious	indolent
persevering	unsteady
independent	independent
jealous of their own liberties, and just to those of others	zealous for their own liberties, but trampling on those of others [i.e. the slaves]
interested	generous
chicaning	candid
superstitious and hypocritical in their religion	without attachment or pretensions to any religion but that of the heart

So Jefferson's theory makes sense to us, and if there is a surprise, it is that his theory sounds so contemporaneous. And, indeed, as I will show below, we shall find that his cognitive theory, in all its particulars, occupies a central place in modern jurisprudence (it is cited by a Supreme Court Justice in 1987), as well as in the political judgments of 1776. Moreover, it is *acted upon* today.

In short, Wills shows us how Jefferson's cognitive theory guided the creation of a government meant to be founded on moral principles rather than on hereditary power and authority. And whereas that theory resembles modern cognitive theory in emphasizing the distinction between intuition and analysis, it is in *opposition* to modern views that it gives the primary value to intuition. It was that very point that allowed Jefferson to believe what he said when he declared that "all men are created equal," for he meant that they were created equal in the "moral sense," in his view a more important cognitive faculty than formal reasoning. And that has a significant bearing on contemporary issues. The most prominent effect of Jefferson's theory can be seen in the cognitive theory of the U.S. Supreme Court which prevailed in the latter part of the twentieth century. The cognitive activity of the Justices of the Supreme Court are of special interest to us because we assume that the Justices are well trained, greatly experienced, and devoted in principle to analytical cognition and the production of coherent moral judgments. Therefore, it will be of considerable interest to us to learn how head and heart are separated—if, indeed, they are separated—at the U.S. Supreme Court.

Separating Head and Heart in Moral Judgments at the U.S. Supreme Court at the End of the Twentieth Century

During the trial of Timothy McVeigh, the man convicted and sentenced to death for bombing the federal building in Oklahoma City in 1995, the judge urged the attorneys not to "inflame or incite the passions of the jury" (Thomas, 1997, p. A1). Those admonitions offer a sharp division between a demand for reason devoid of emotion and reason based on passion (discussed in chapter 2). And they reflect exactly what we would have expected the judge to say. However, just as psychologists are not unified on the matter of separating emotion and reason, neither are the courts. Indeed, we shall find that even in the highest courts, the *passionate* application of justice is urged on us by some of our most respected jurists while denied us by other equally respected jurists as offending our basic principle of "a government of laws not men." We shall see that those who insisted on the presence of passion in the application of justice have been praised for the actions which have followed from their exhortations and have significantly affected our social policies. But we shall see that the opposite—careful allegiance to rationality and reason—has also occurred. And to that situation we now turn.

INTUITIVE COGNITION AT THE U.S. SUPREME COURT

It is a long leap from Jefferson to Oliver Wendell Holmes, Jr. (1841–1935), but it is a leap required by our topic. Holmes is a good place to stop, however, because he was one of the most influential justices of the Supreme Court. I pause here only to note his most famous sentence: "The life of the law has not been logic; it has been experience" (Holmes, 1923). (I have elaborated on the importance of this sentence to cognitive psychology elsewhere; Hammond, 1996). A similar distinction between analysis and intuition appears in the biography of Earl Warren, Chief Justice of the U.S. Supreme Court from 1953 to 1969. According to Warren's biographer, Ed Cray, one of Warren's clerks, Michael Heynman, noted that "weighing cases, he reacted to *personal experience* (italics added) rather than philosophical abstraction or contemplation. Relying on well-understood personal values, he reached decisions quickly, more rapidly than most of the brethren" (Cray, 1997, p. 356). Rapid decision making, based on personal experience, is a widely acknowledged sign of intuitive cognition (see below). And a second clerk, Dallin Oakes, who often disagreed with the Chief's indifference to legal niceties, said to Cray that "he went for a decision that would be right, and right for him was what was morally based, what was good for the individual, and in the larger sense good for the country to have a fair society" (p. 356).

But his intuitive judgments occasionally produced surprises, as intuitive judgments will. For in a case involving the burning of the flag, Warren surprised even his clerks by voting with the most conservative judges to make this act unlawful speech, not simply symbolic speech, as he would have been expected to do. Cray reports that "at lunch with his clerks the following day, one of them asked, 'What's the theory of your dissent?' There was a long pause before Warren replied. 'Boys, it's the American flag. I'm just not going to vote in favor of burning the American flag' " (p. 492). No legal principles here; simply a feeling of what was "right." This response may have been what Justice Holmes had in mind many years earlier when he indicated "the life of the law has not been logic; it has been experience."

This approach to constitutional law continued past Warren to the more sophisticated opinions of Justice William Brennan (1906–1997). It can be found in an annual Benjamin Cardozo lecture delivered in 1987 by Justice Brennan to a meeting of the Association of the Bar of the City of New York. The title of Justice Brennan's lecture was directly related to our topic: "Reason, Passion, and the Progress of the Law" (Brennan, 1987). Brennan was forthright in his exposition of and belief in the principles set forth by Cardozo, a Supreme Court Justice famous for his liberal decisions during the Roosevelt era, the essential component of which was that passion—the heart—did have a place, indeed, perhaps *more* of a place in the administration of justice than the head. Brennan pointed out that Cardozo "attacked the myth that judges were oracles of pure reason, and insisted that we consider the role that human experience, emotion, and passion play in the judicial process," a remark highly reminiscent of Holmes's comments. And, Brennan noted, Caradozo's views are still with us for "his lectures provoked a lively controversy whose implications preoccupy us today" (1987, p. 953). Brennan, also well known for his liberal opinions on the Court (1956–1990), was equally sensitive to the contemporary nature of the struggle between head and heart on the bench. His concern with that struggle can be seen in his remark that "in our own time, attention to experience may signal that the greatest threat to due process principles is formal reason severed from the insights of passion" (p. 968).

But shouldn't "formal reason *be* severed from the insights of passion"? Isn't that what we have been taught to expect from the solemn administration of justice? Isn't that what the judge in the Oklahoma City bombing trial wanted when he insisted that the attorneys in the case were not to "inflame or incite the passions of the jury"? And isn't that why we are so interested in reducing or eliminating the putative deleterious effects of stress on judgment? But eliminating those effects seem to be exactly opposite to what Justice Brennan wants in the administration of justice, for Brennan thinks that "the greatest threat to due process principles is formal reason *severed* from the insights of passion." Therefore, it follows that he wants "formal reasoning *unsevered* from the insights of passion" developed by the jurors in the Oklahoma City bombing trial. He wants them to *include* whatever "insights" their passion might provide. One wonders, does he want the victims' family members sitting in the courtroom also to express whatever "insights" their passions might provide *them*? Does Justice Brennan wish to be sympathetic to the perpetrator of the bombing, who has already expressed in deeds the "insights" that his passion provided *him*?

All of which makes us ask: What did Justice Brennan mean by his use of the term "passion"? Fortunately, he tells us, and tells us that he chose the word "passion" deliberately; it is "a word I choose because it is general and conveys much of what seems at first blush to be the very enemy of reason. By 'passion' I mean the range of emotional or *intuitive* (italics added) responses to a given set of facts or arguments, responses which often speed into our consciousness far ahead of the lumbering syllogisms of reason" (p. 958) No student of judgment and decision making would quarrel with that concept of intuitive responses! (Nor, we can suppose, would Chief Justice Warren.) Very similar definitions of intuition can be found in a variety of books on judgment and decision making (cf. Hammond, 1996). There is nothing wrong with Justice Brennan's concept of intuitive judgments; he has it right. The question is: Is that the sort of cognitive activity to be desired in the even-handed administration of justice? Do we want Justice Brennan, or any other Justice, to decide his cases on the basis of "responses which often speed into our consciousness"? Or do we prefer slow, deliberate, thoughtful analytical work on the part of the judge?

The answers to these questions are all too easy. Even Brennan recognizes the dangers of allowing passion to participate in the judge's cognitive activity:

> The judge's job is not to yield to the visceral temptation to help prosecute the criminal, but to preserve the values and guarantees of our system of criminal justice. . . . Indeed, the judge who is aware of the inevitable interaction of reason and passion . . . is the judge least likely in such situations to sacrifice principle to spasmodic sentiment. (Brennan, 1987, p. 961)

Of course we want the judge to be analytical; of course we want the attorneys in the Oklahoma City bombing trial to follow the judge's instructions not to incite the "passions" of the jurors. Indeed, we want the *judge* to avoid passion in his judgments. Then how do we explain Justice Brennan's enthusiasm for the "inevitable interaction of reason and passion"? Astonishingly, Brennan makes it clear that their foundation lies in Thomas Jefferson's cognitive theory that separates "Head" from "Heart." For Brennan notes that "two hundred years ago, these responses would have been called responses of the heart rather than the head" (as I have shown to the reader in the pages above). And to make unmistakably clear that he is conscious of the dual nature of cognition that Jefferson expressed two hundred years earlier, Brennan states: "To individuals such as Thomas Jefferson, the faculty of reason was suited to address only questions of fact or science, while questions of moral judgment were best resolved by a special moral sense, ***different from reason*** (italics added), and often referred to as the "heart" (pp. 958–959). In his letter to Maria Cosway, Jefferson stated that "morals were too essential to the happiness of man to be risked on the incertain [sic] combinations of the head. [Nature] laid their foundation therefore in sentiment, not in science" (Jefferson, 1984, p. 874). Indeed, Jefferson could have cited Pascal (1602–1674) who observed that "the heart has reasons that reason does not know of."

So we see that Brennan could not forego an appeal to the "heart," to the "moral sense different from reason." Therefore, when Justice Brennan moved beyond the purely legal—and thus purely analytical—issue of constitutional compatibility, he moved from the analytical pole of the cognitive continuum in the direction of the intuitive pole of the continuum to the central region of *quasi* rationality. His enthusiastic decision to include passion as part of his cognitive activity meant that he moved from purely analytically derived judgment to moral judgment, and, most important, to a "moral sense different from reason." For a Justice of the Supreme Court to acknowledge that his or her judgment would be based on a process "different from reason" would of course require an explanation; therefore, a defense for a departure from reason must be forthcoming. How will such moral judgment be defended, now that his cognitive activity has been expanded to include a process "different from reason"? How will Justice Brennan defend his move to a quasi-rational cognitive process at the judicial branch of government—the Supreme Court—that is most committed to full rationality? Once he has given up analytical *coherence* within the framework of the Constitution as a defense for his judgment, and once he has decided to rely on passion as well as reason and rationality, Justice Brennan must find a defense for his judgment that he admits departs from that standard. He must defend his moral judgment somehow. What course will he take?

We know what the answer will be. He will seek justification by reference to the correspondence of his judgments to those of his peers (as discussed above). And Justice Brennan does just that; there is nowhere else to go; he reaches back to Cardozo's views on the source of the difficulties of exercising one's judgment on the Supreme Court. He notes that Cardozo laid these difficulties at the door of *uncertainty:* "Judging is fraught with uncertainty. Principles and logic take a conscientious judge only so far; then, in Cardozo's words 'the paths. . . . begin to diverge, and we must make a choice between them' " (1987, p. 952). Apparently, Brennan approves of Cardozo's belief that logic might well take the judge in one direction while his "principles," his moral judgments, might point in another. He then quoted Cardozo at length in his attempt to sweep away the "myth" that judges' thoughts are somehow lifted "into the realm of pure reason."

In short, Brennan sought justification for his departure from reason (and thus legal coherence) by seeking correspondence with judgments by his peers (another judge's departure—in this case Cardozo's—from reason). That is because intuitive, or quasi-rational, judgments, need correspondence with other judgments similarly derived for their justification. Analytically based judgments, however, need to be coherent in detail with other rational, analytically based judgments, irrespective of their correspondence or agreement with the passionate judgments of their peers.

Thus, we see that two hundred years after its introduction, Jefferson's theory of cognition—that included moral judgments—was not merely acknowledged as historical fact, it was heartily appropriated, made to be explicit, and made to be an essential and desirable element of judicial decision making. (I discuss the disagreements in moral judgments made at the Supreme Court by Justices Blackmun and Scalia over the constitutionality of the death penalty in Hammond, 1996, pp. 343–350.)

An interesting parallel in the application of the distinction between the justification of moral judgments in terms of coherence or correspondence can be found in an unexpected place—the letters of Civil War soldiers to their families. Thousands of these letters were collected by James McPherson (1997) in a book aptly titled *For Cause and Comrades: Why Men Fought in the Civil War.* McPherson shows that soldiers often wrote to their families about their dedication to the cause of "preserving the union" (in the North) and to the cause of "freedom and liberty" (in the South), thus justifying their absence from home and hearth in terms of a "cause," an overriding, and to them, coherent, idea that they could not abandon. In addition, however, soldiers on both sides, time and again, explained how they cared nothing for the grand ideas of those in power, but that they could not stand to be separated from their comrades, could not stand to "let them down," and that it was their close link to their comrades that kept them coming back, thus justifying their absence on the correspondence of their moral judgments with those of their peers rather than the coherence of an idea.

Thus, my answer to the question of whether the conceptual framework of this book can be extended so as to aid in the understanding of moral judgments is "yes." I reviewed the concept of constancy as a theory of stress, extended its application to moral judgments, extended the application of the metatheories of coherence and correspondence from professional-technical judgments to moral judgments, using the distinctions offered by James Q. Wilson, and applied the concepts of intuition and analysis from Cognitive Continuum Theory to the theory of cognition employed at the Supreme Court by Justice Brennan. I acknowledge that these extensions are not based on experimental evidence, and therefore they must be considered illustrations—valuable illustrations, I hope—rather than proof. Nevertheless, I consider the theory to be made sufficiently clear through these illustrations to make them testable and thus falsifiable.

Although I have indicated how this conceptual framework can be applied to various forms of moral judgment, I have not yet met the challenge of applying it to moral judgments under disruption of constancy, a topic to which I now turn.

Disruptions to the Constancy of Moral Judgments

The reader will recall that earlier I distinguished between two types of disruption, endogenous and exogenous. Endogenous disruptions occur *within* the environmental task system the person is coping with (e.g., the operating system of an aircraft or computer). Exogenous disruptions are *external* to the operating system (e.g., fire in the lavatory of a plane) in the situation in which the person is acting. Of course, these disruptions do not always maintain their independence; what began as an exogenous disruption may develop into an endogenous one (an external fire may develop into a threat to the internal mechanisms). In what follows I show how this distinction applies to moral judgments, as well as to professional-technical judgments.

Endogenous Disruption to Coherent Moral Systems

As noted above, endogenous disruption—within-system disruption—is most threatening to moral systems that depend on coherence for their legitimacy. Thus, a disruption to what had been assumed to be a coherent argument may cause a moral judgment to fail. Examples of endogenous disruptions include declarations by an opponent of the discovery of an inconsistency in one's moral system, or the necessity of a public declaration by oneself of having discovered an inconsistency in one's own moral system.

We are all familiar with both kinds of endogenous disruptions. The first occurs during an argument with an opponent who successfully points out that you have said one thing on one occasion and the opposite on a second. This creates a loss of constancy in a moral system that had claimed coherence and now must face up to the fact that it does not. No one has to be told that such loss of constancy is unwelcome.

Discovering one's own incoherence may be even more difficult. This situation occurs because moral judgments rarely present themselves singly; almost always we are faced with choosing between accepting one value and rejecting, or at least, relinquishing, another deeply held value or values. Such circumstances—Patrick Henry faced them—place strong, often impossible-to-meet, demands on the coherence of our moral principles. The classics—for example, the morality plays—devote themselves to this eternal feature of human existence.

The abortion debate is a current example of two strong values—the life of a fetus or unborn child is pitted against, perhaps, the life of the mother—in a bitter contest. And often the contest is posed as the defense of one "right" against the denial of that "right." As we all know, there will be those who choose one value and will not only relinquish the other(s), but also will deny their validity, and when opposition occurs, accuse their opponents of advocating inhumanity, deceit, and worse. Regrettably, our lack of knowledge about how to resolve such disputes, if, indeed they can be resolved, leaves everyone subject to endless tirades that merely repeat themselves year after year. Even more regrettably, it may be true—and a figure no less imposing than Isaiah Berlin, one of the world's greatest historians of ideas, claims it is true—that, in principle, such entangled and opposing values *cannot* be reconciled (see, e.g., Berlin, 1997).

A competing value operates much like an endogenous disruption in an organized system such as a navigation system, or power plant operation, in which two diametrically opposed courses of action suddenly demand attention. That is parallel to two opposing values demanding action within a heretofore coherent system. I refer to this as an "endogenous" disruption because the opposing actions, or values, occur within the *same* system.

Exogenous Disruptions to Coherent Moral Systems

Coherent moral systems are generally impervious to exogenous disruptions such as attacks on the system that do not penetrate to the interior relations within the system, but simply attack the system as a whole. Thus, for example, defenders of what is sometimes called the Judeo-Christian code are apt to ignore attacks on that code that are general and do not single out specific "commandments" for criticism. Similarly, broad attacks on general moral systems because of the immoral behavior of their constituents have little impact. As long as the moral system can maintain the internal integrity of its argument, such attacks are fruitless. And just as the pilots of the Aloha flight that lost part of its fuselage survived because they ignored this exogenous attack on their aircraft and continued to carry out their task of flying the airplane, so too do the proponents of moral systems survive exogenous attacks on their endeavors—by ignoring them. This sharply contrasts with the response to an endogenous attack on the internal features of the system, for in that case, change is more likely to occur.

Endogenous Disruptions to Correspondence Judgments of Morality

Correspondence judgments are not vulnerable to endogenous disruptions because they make no claim to rationality; they are simply built on "inalienable feelings," in Voltaire's words, on intuitive judgments that are justified by support from one's peers who seldom, if ever, demand logical justification.

Bar-Hillel and Yaari address this situation in their empirical studies of judgments of fairness. They emphasize the need for objective standards with which to compare subjective judgments about justice, and point out the useful role such standards have played in the discovery of the constancies in perceptual judgments. Size constancy, for example, could not have been discovered were it not for the fact that we have physical measurements of size to use as standards for the accuracy of judgments (see the discussion of constancy in chapter 6). But they do not believe that it will be possible to establish similar objective standards with which to compare intuitive judgments of justice. As Bar-Hillel and Yaari put it,

> It is far more likely, however, that "soft" constructs such as morality or beauty will prove ultimately to be in the eyes of the beholder, in the sense that a theory of these constructs will have no independent standing from a theory of judgments thereof. Such also, we believe, is the construct of distributive justice. (pp. 58–59)

If we take that statement as truth (as I do), it means that it will only be possible to compare one person's *intuitions* about justice with intuitions of another. But intuitions are entirely subjective. That is what is meant by saying that justice is "in the eye of the beholder." It will not be possible to ascertain which person's judgment was truly just—that is, most closely approximates an objective standard—and which person's was unjust—was the farthest from the standard—for there is no independent standard by which such a conclusion can be drawn.

Thus, there is little chance for a reasoned discussion to overwhelm moral judgments based on correspondence with those of one's peers. Since correspondence judgments of morality do not demand coherence, disruptions to the consistency of correspondence judgments are therefore not fatal; indeed, they are irrelevant to the judgment process. Persons making correspondence judgments are focusing on the behavior of the objects of their judgments, not the consistency of their judgments, therefore consistency is not a goal. The goal is to make the "correct" judgment, to declare certain behaviors as right or wrong, acceptable or not acceptable, where "correct" is generally defined by community mores or agreement with peer judgments. The rationality of such judgments is of no concern.

Exogenous Disruptions to Correspondence Judgments of Morality

Because correspondence judgments of morality focus on making the correct judgments of behavior, exogenous disruptions will be those that disrupt, or deny the validity of, the *entire* system. These disruptions are apt to be as ineffectual as they are severe. They are likely to come from persons with an entirely different sociocultural background, those who constitute a different set of peers, and therefore persons who have no inclination to change their moral systems, nor to develop an appreciation of another's. But they tend not to be any more effective than endogenous disruptions.

The reason for lack of change lies in the absence of cross-cultural standards. Absence of a standard allows injustice, more broadly, evil itself, to escape from objectivity and thus to appear only in the eye of the beholder. History is filled with the horrors that illustrate that conclusion; firm believers in the sanctity of the Bible and Ten Commandments, for example, have throughout history believed it entirely proper to torture and slaughter their opponents, and still do, just as staunch believers in the Koran happily and enthusiastically murder "infidels" who "deserve" their fate. The attempted extermination of the Jews by the Germans is the most notorious example of self-righteous villainy. Note that in these cases it is their unshakable belief in the "rightness" of their moral code that is driving the torturers and murderers to their unspeakable deeds. In the eyes of these beholders justice is being done, and the purveyors of that justice hope to establish for themselves a firm place beside the throne in the kingdom of heaven. And so on. Without an objective standard the exterminators need turn only to confirmation by their peers.

Extension of the Theory

Does it really matter whether the theory gets extended? I believe it does because I believe it can enlighten even moral philosophers. Here is an example of Isaiah Berlin, unconsciously sliding from moral judgments within the coherence metatheory to moral judgments within the correspondence metatheory, thus rendering his conclusion less valuable. Note how he recognizes the impossibility of reconciling different moral judgments within the coherence metatheory, and as a result, calls for *compromise,* without recognizing that compromise is only possible within the correspondence view.

> If the old perennial belief in the possibility of realising ultimate harmony is a fallacy, and the thinkers I have appealed to—Machiavelli, Vico, Herder, Herzen—are valid, then if we allow that Great Goods can collide, that some of them cannot live together, even though others can—in short, that one cannot have everything, in principle as well as in practice—and if human creativity may depend upon a variety of mutually exclusive choices: then, as Chernyshevsky and Lenin once asked, "What is to be done?" How do we choose between possibilities? What and how much must we sacrifice to what? There is, it seems to me, no clear reply. But the collisions, even if they cannot be avoided, can be softened. *Claims can be balanced, compromises can be reached.* (Berlin, 1991, p. 17; italics added)

But compromise can only be reached via a correspondence metatheory, coherence theories must be taken on an all or nothing basis; otherwise, they lose their most important asset, namely, coherence. And that means a return to Voltaire (and James Q. Wilson), and that, in turn, means a considerable amount of cognitive baggage, such as intuitive judgments, must come along as well. We can see this as Berlin becomes specific:

> In concrete situations not every claim is of equal force—so much liberty and so much equality; so much for sharp moral condemnation, and so much for understanding a given human situation; so much for the full force of the law, and so much for the prerogative of mercy; for feeding the hungry, clothing the naked, healing the sick, sheltering the homeless. Priorities, never final and absolute, must be established. (p.17).

The reader will recognize that Berlin is, unwittingly, calling for a correspondence metatheory to be applied to moral judgments. And the reader will also recognize (Did Berlin?) that every correspondence theory will bring problems of its own, as indicated above.

So the reader cannot expect a final, all-encompassing answer from me—or anyone else. Many have tried to provide an answer that will be universally accepted but all, including Berlin, have failed. Mine was a more modest goal; simply to extend the conceptual framework developed in relation to professional-technical judgments to the area of moral judgments.

Is it possible that I have achieved that? The opening paragraph of this chapter taken from Earle indicated that a psychological analysis of moral judgments was impossible because a psychological analysis would assume that the individual was not a free agent, but was controlled by psychological determiners of behavior. If this is true, then the individual is not free, and therefore his or her moral judgments are "no longer properly moral behavior" for "an agent can act morally only on the condition that it is free." So it is fair to ask in light of this argument (which I do not dispute) "in what sense have you [the author] increased our understanding of moral judgments"?

I call the readers' attention to the fact that I have laid down no laws of behavior that putatively control human judgment; I see no evidence that such judgments are lawfully controlled, and I do not pretend that I have offered any. I do not take the approach of a physicist to the topic of human judgment. And I do not believe the study of human judgment at the time of this writing is different from other fields of psychology. I know that during the first half of the twentieth century psychologists dreamed, hoped, and expected that their science would produce laws of behavior (see my remarks in the Preface regarding the "devastating review" provided by the volumes edited by Sigmund Koch). But those days are over or so it seems to me.

Thus, my efforts are more like those of a geographer than those of a physicist; I try to describe forms of human judgment evoked or induced by environmental circumstances. I offer a conceptual framework that provides a description and explanation of human judgment enacted by a free human agent, and the conditions that induce various forms of it, and thus increases our understanding of moral behavior without denying the antinomy pointed out by Earle. In short, without positing psychological laws that would mitigate moral freedom, I have provided the reader with a coherent theory that intends to describe moral judgments and correspond to the facts of behavior.

10

Conclusion

Current theories of the effects of "stress" on judgment (or other behavior) are unsatisfactory for a number of reasons: Many are impossible to falsify, and they all have an escape route for negative results. And the escape route is always the same; it lies in the expansion, contraction, or other revision in the definitions of stress. Revisions are not objectionable in themselves; indeed, one is offered here. But these are generally open-ended, largely because of the recourse to subjective expressions of stress. That is, for purely psychological theories, we only know when and if "stress" occurs if the subject tells us that it has or has not. If that conclusion is correct, then it is imperative to develop a falsifiable theory of the phenomena of interest without the baggage of the term "stress" or the "perception of stress." That will require independent definitions and determination of both environmental and subjective conditions, and, we can hope, contributions from neuroscience.

The theory described here does have the important advantage of being falsifiable. It rests on the long empirically established phenomenon of constancy, a bedrock principle of a century of research, so central to psychology its importance has almost been forgotten. The theory takes the novel step of differentiating the phenomenon of constancy into two forms, correspondence constancy and coherence constancy, thus drawing on metatheories that differentiate the judgment process into its two essential forms, each of which has received considerable research attention reflected in hundreds if not thousands of research articles. Regrettably, few, if any, of these studies have been identified as relevant to or representative of these two metatheories, and as a result, there is much confusion about the relation of these numerous studies to one another.

The present approach differs from convention in that it rests on a theory of task conditions, as well as a theory of cognitive functions; that is important because both theories are needed to establish generality of conclusions. Moreover, separate theories are needed to avoid the circularity inherent in the use of subjective definitions of "stress." Thus, the present theory emphasizes the need for careful analysis of the change in task properties that occurs when an unexpected disruption occurs. Indeed, task properties are precisely the locus of the unexpected disruption. The task properties at t_1(prior to the time the unexpected occurs) *and* the task properties of the new situation at t_2 (the time the unexpected occurs) demand description in theoretical terms so that hypotheses can be generated and research can put them to the test. All of the predictions and hypotheses I have suggested rest on the five premises put forward earlier.

The psychological concept of constancy and the disruption of constancy—in any of its forms—are the source of what is now vaguely labeled "stress." The theory is specific about the threats to constancy from disruption and destabilization. Threats arise from two main types of task events—endogenous (within task events) disruption of constancy, and exogenous (extratask events) disruption of constancy. Each type of disruption is expected to produce different cognitive activity depending on the initial and final location of that activity on the cognitive continuum, and depending as well on the location of the task—and changes in the task—on the Task Continuum Index. These predictions are made specific in the diagrams earlier (see Figures 8.1 and 8.2). In short, I intend to show how a theory of what the layperson calls judgment under "stress" need be no more than a theory of task-cognition interaction rooted in the well-documented phenomena of constancy.

Finally, in what is clearly a risky effort, I include a chapter in which I extend the general framework applied to professional-technical judgments to moral judgments. Chapter 9 addresses the question of whether the conceptual framework can be systematically applied to the "soft" material of moral judgments. I believe this demonstration shows that it can.

New Orientation Offered by Theory

A new theory should offer a new orientation, a new way of organizing the basic materials of the field, and should open new research approaches, particularly in areas that seem to be stalled, such as stress research. The conceptual framework offered here does that in several ways, all of which can be summarized as follows: *(1) Both environmental events and cogni-*

tive activities must be considered jointly, with equal attention to both; (2) each should be more differentiated than they are at present; and (3) their interaction should be the center of our attention. Some examples follow.

1. *"Stressors" should always be examined in relation to cognitive activity.* Time pressure is generally assumed to be a stressor under any circumstances, but is it? Only if the operating cognitive process is ignored. The theory outlined above, however, indicates that it should never be (see Figures 8.1 and 8.2). More specifically, the theory asserts that one of the defining properties of intuitive cognition is its rapidity of process. Thus, rapid judgments are inherent in intuitive cognition, and, therefore, time restrictions should have minimal, if any, effects on intuitive cognition. Time pressure can be a stressor only in connection with *analytical* cognition, a process that always requires time for its execution. The smoke jumpers running from the fire were not under time pressure for their intuitive judgment to form. The foreman was under time pressure for his analytical cognitive activity to reach coherence, however. Therefore, it is only when analytical cognition is demanded by the task that time pressure becomes a threat to successful cognition, and then it is a threat only to the analytical demands of coherence constancy. The foreman's creativity under time pressure, as well as other stressors, was indeed an achievement.

In short, time pressure is a significant threat to successful cognition only in Cell 9, Figure 8.1. It follows that time pressure is less of a threat to quasi rationality (Cell 5), because quasi rationality demands less analytical work than Cell 9. Time pressure therefore is not a threat at all in Cell 1.

The theory thus differentiates specific conditions, such as time pressure, among those that will and will not be "stressors." In this way the theory moves us away from folklore; it prevents us from merely taking for granted conditions that *sometimes* are "stressors" and extrapolating them to all circumstances.

2. *Common disruptions of constancy should be differentiated in terms of (1) whether they are exogenous or endogenous and (2) which cognitive mode is in operation.* The stressors commonly used by stress researchers include heat, vibration, noise, and similar exogenous events. But they are derivatives from folklore, considered in very general terms, and force researchers to rely on the readers' experience with such events to appreciate their significance; their differentiation into endogenous or exogenous events is seldom considered. Never considered is the question of which cognitive mode is operative at the time of the stressor's impact. Both types of differentiation are important, however.

For example, when fire is an *exogenous* event that impacts an *analytically coherent* operating system, the proper, and natural, cognitive response is to continue functioning in that mode. Thus, if fire occurs in the lavatory of an airliner, it is an exogenous event, and the pilots will continue their analytical work. Fire in an engine, however, is a different matter; it is an *endogenous* event that disrupts the operating system from within. Under these circumstances a shift in cognition to a more intuitive mode would be induced (as it was induced in the case of Captain Haynes). But system designers realize both the propensity for fires to occur in engines *and* for humans to shift to the intuitive mode when fires occur. As a result, fires are not unanticipated; the analytically most appropriate technical response is constructed, and the analytically most appropriate cognitive response is constructed through training. (Captain Haynes had neither of these advantages.)

The necessity for constant vigilance against an *incorrect* cognitive shift can be seen in the frequent and regular retraining of pilots to use the technically constructed apparatus and the analytical mode of cognition to operate it. Such training offers testimony to the need to defend against the natural tendency to shift to an intuitive mode of cognition under these circumstances. In short, classifying the event "fire" (or heat, noise, or vibration) as a "stressor" does not help us, and indeed, may very well mislead us. What does help us is a theory of environmental and cognitive conditions that shows how the two sets of concepts interact.

3. *Leaders and followers should be taught to accept cognitive change.* In all likelihood no one gave Commander Rinn or Foreman Dodge lessons about cognitive change, but they accomplished it, either because of, or in spite of, enormously difficult circumstances. And their changes, although different from one another, had a marked effect on their successes. How often others do this with what results is unknown. But if we want more leaders like these two men, in our administrative and executive offices, as well as in the face of physical life-threatening circumstances, we need to discover whether such changes are always or usually effective, whether teaching about such change is in fact necessary, and if so, whether it is feasible. The above examples indicate the putative value of such teaching, but we have no idea whether it is feasible because it has never been tried. From this point of view, the goal of training is not merely to produce skill in performing certain actions by turning unfamiliar activities into routine activities through repetition. It must also be to teach the crew to have confidence in the leader when his or her cognitive activity is forced to be *different from what it is expected to be*, which will often be the case under conditions of disrupted constancy.

The idea of cognitive change has been approached many times in the history of psychology, but as yet there has been no systematic effort to study change in cognitive *process*, in contrast to content. Nor have the consequences for leaders and followers been examined. Nevertheless, as our examples show, leaders *will* engage in dynamic cognitive activity of one form or another, whether they plan to or not, whether they recognize it or not, when faced with unanticipated conditions that disrupt coherence constancy or correspondence constancy. Research should teach us more about how to prepare both leaders and followers for the implications of such changes, as well as how to anticipate the differential effects of endogenous and exogenous disruptions and their consequences.

Perception of change in cognitive function will be particularly important in the case of moral judgments, if the theory can successfully be extended to them. For, as the theory argues, a change from one metatheory to the other means a change in the criteria for the validity of the judgment, and, of course, it will be important for disputants to recognize this.

I offer these brief examples of a new orientation, suggested by the theory outlined here, principally to show that the theory does point us in new directions.

11

Appendix: Literature Review

Note: In what follows the term "stress" is used in the loose sense in which it has been traditionally used colloquially and in the stress literature.

Progress in the field of stress research is marked by a clear division of three literatures (indicated also in a history provided by Appley & Turnbull, 1986). One literature is oriented toward the clinical/personality/social psychology of the effects of stress on cognition and other processes, exemplified by the persistent work of Lazarus, Janis, and others. A second literature is oriented toward ergonomics, or human factors psychology, exemplified by the work of Hockey, Hamilton, and Hancock among others. As might be expected, the former is dominated by theory, research, and generalizations that are broader in scope, less precise in terminology and measurement, less experimental and more psychological-test related than the latter. Both literatures are voluminous, but the former has far more popular appeal than the latter. Also, as might be expected, the two literatures exist in almost perfect isolation; cross-citations are almost nonexistent. The second, human factors literature, does have some links with the third, psychophysiological literature, but none to the fourth, the judgment and decision making literature.

Most important for the purpose of this monograph is the fact that these three literatures do not include any significant amount of material from modern research (1960–1990) on judgment and decision making. There are a few scattered, brief references to work by Kahneman and Tversky, but the work of Norman Anderson, Hal Arkes, Berndt Brehmer, Robyn Dawes, Michael Doherty, Ward Edwards, Hillel Einhorn, Baruch Fischhoff, Reid Hastie, Robin Hogarth, John Payne, Paul Slovic, and

Thomas Wallsten, to mention only a few, is essentially ignored. Either the thousands of theoretical and empirical research articles in the contemporary literature of judgment and decision making (J/DM) have not been read, or have been read and judged to be irrelevant by the researchers in the three fields mentioned.

Only recently have human factors psychologists begun to notice the J/DM literature. As noted above, a recent book by Hancock entitled *Human Factors Psychology* (1987) includes a chapter by a judgment and decision researcher (D. Kleinmuntz) that describes this literature, and a chapter on short-term memory by Klapp that cites recent work in cognitive psychology. But the other chapters in this book generally ignore this work. Even Hancock's chapter (with Chignell) that includes a section on "Stress and Adaptive Functioning," scarcely alludes to modern judgment and decision research, and makes only one passing reference to the clinical/personality/social literature.

It seems reasonable to suppose that our understanding of this complex topic would be enhanced if the four literatures were brought into contact with one another. Therefore, in Literature I, I try to organize them in a manner that will prove useful to those who are producing work within the judgment and decision making field.

1. Literature I: Clinical/Personality/Social Psychology

Theories

Janis, Lazarus, and Mandler are among the prominent theorists in the clinical/personality/social literature of stress and cognition. However useful their theoretical work may be for the purposes the authors intended, it does not provide a degree of formality sufficient for the type of research on judgment and decision making considered here. Also, as might be expected under these circumstances, the theories within this literature do not compete. Indeed, the isolation *between* Literatures I, II, and III is paralleled by isolation *within* Literature I; the theorists hardly acknowledge the existence of other theorists. For example, in the recent far-ranging article entitled "Theory-Based Stress Measurement," Lazarus (1990) does not cite Janis, Mandler, or *any* of the theorists mentioned below. The reverse is also generally true; they seldom cite him. In short, a primitive, idiosyncratic state of theorizing exists in Literature I; such theories apparently have considerable popular appeal but do not offer cumulative science.

Nevertheless, current theories of stress in Literature I—which generally encompass a much wider range of psychological functions than judgment and decision making—offer numerous ideas and hypotheses that can hardly be ignored by researchers in this field, and that almost certainly would be reinvented if they were ignored. Therefore, I present brief descriptions of several theories that bear on the topic of stress and its cognitive consequences. Regrettably, they must simply be noted and briefly characterized; comparison is impossible because of their differences in form and content.

In order to avoid miscommunication, insofar as possible, direct quotations from the author's text are presented and commented on briefly.

Lazarus

The following quotations from Lazarus's article on "Theory-Based Stress Measurement" will indicate his present views:

> Psychological stress refers to a particular kind of relationship between person and environment (Lazarus, 1966; Lazarus & Folkman, 1984, 1987). The stress relationship is one in which demands tax or exceed the person's resources. The unit of analysis is an ongoing transaction or encounter which is *appraised* by the person as involving harm, the threat of harm, or a positive, optimistic, mobilized, and eager attitude about overcoming obstacles, which I have called *challenge*. Once a person has appraised a transaction as stressful, *coping* processes are brought into play to manage the troubled person-environment relationship, and these processes influence the person's subsequent appraisal and hence the kind and intensity of the stress reaction. This cognitive-relational view, which once had to overcome entrenched behavioristic resistance, is now all but dominant. There is no need, therefore, to recount in detail the concepts of appraisal and coping, which have now become part of the routine vocabulary of researchers in the field of stress.
>
> *Transaction* implies that stress is neither in the environmental input nor in the person, but reflects the conjunction of a person with certain motives and beliefs (personal agendas, as it were) with an environment whose characteristics pose harm, threats, or challenges depending on these person characteristics.
>
> Transaction also implies *process*. The stress relationship is not static but is constantly changing as a result of the continual interplay between the person and the environment. For example, in problem-focused coping, the actual terms of the relationship are changed, which in turn affects the appraisal. In emotion-focused coping, what is attended to may be changed, or its meaning is changed as when the person denies or distances from the threat, which in turn also affects the appraisal. In effect, stress is a multivariate process involving inputs, outputs, and the mediating activities of appraisal and coping; there is constant feedback from ongoing events, based on changes in the person-environment relationship, how it is coped with and, therefore, appraised (see Folkman & Lazarus, 1988).
>
> This view has dramatic consequences and poses great difficulties for stress measurement. It abandons a simple input-output analysis and becomes a fluid systems analysis involving a host of variables that influence each other in time and across the changing contexts of adaptation. The best way for the reader to see this is to examine the system of interde-

pendent variables and processes in Table 1, which is one version, and to ask the question of where and what stress is, in this kind of analysis, and how it might be measured. (pp. 3–4)

These remarks should be sufficient to enable the reader to grasp the conceptual level of Lazarus's theory of the effects of stress on behavior. They will also serve to enable the reader to see the distance between theorizing of this type and that developed in the field of judgment and decision making.

Mann and Janis (1982)

In this article, Mann and Janis present a "conflict theory" of decision making. Their model "is primarily concerned with identifying factors that determine the major modes of resolving conflicts. It describes how the psychological stress of decisional conflict affects the ways in which people go about making their choices" (p. 341). Mann and Janis "postulate that there are five basic patterns of coping with challenges that are capable of generating stress by posing agonizingly difficult choices" (p. 344).

The first coping pattern is "unconflicted adherence [in which] the decision maker complacently decides to continue whatever he or she has been doing, which may involve discounting information about risk of

Table S1.1 Illustrative System Variables for the Stress and Emotion Process

Causal Antecedent	Mediating Process	Immediate Effect	Long-Term Effect
Person variables Values, commitments, and goals General beliefs, such as: Self-esteem Mastery Sense of control	Encounter 1 . . . 2 . . . 3 . . . n Within an encounter Time 1 . . . 2 . . . 3 . . . n Primary appraisal (stakes)	Affect	Psychological well-being
Being Interpersonal trust Existential beliefs	Secondary appraisal (coping options)	Physiological changes	Somatic
Environmental variables Demands Resources (e.g., social support network) Constraints Temporal aspects	Coping (including use of social support) Problem-focused forms Emotion-focused forms	Quality of encounter outcome	Social functioning

Note: Although not shown here, the model is recursive. Also, note parallelism between short-term and long-term effects.

Source: Lazarus, 1990, pp. 3–4. Reprinted here by permission of Lawrence Erlbaum Associates, Inc., and the author.

losses" (p. 344). The second pattern is "unconflicted change [in which] the decision maker uncritically adopts whichever new course of action is most salient or most strongly recommended" (p. 344). The third coping pattern is "defensive avoidance [whereby] the decision maker escapes the conflict by procrastinating, shifting responsibility to someone else, or constructing wishful rationalizations to bolster the least objectionable alternative, remaining selectively inattentive to corrective information" (p. 344). The fourth pattern is "hypervigilance [in which] the decision maker searches frantically for a way out of the dilemma and impulsively seizes upon a hastily contrived solution that seems to promise immediate relief. . . . In its most extreme form, hypervigilance is known as 'panic' " (p. 344). The fifth coping pattern is "vigilance [whereby] the decision maker searches painstakingly for relevant information, assimilates information in an unbiased manner, and appraises alternatives carefully before making a choice" (p. 345). Mann and Janis hypothesize that, in general, the first four coping patterns tend to be maladaptive, whereas the fifth pattern usually meets "the main criteria for high-quality decision making" (p. 345). Mann and Janis also describe various related aspects of conflict theory.

Mann and Janis's "conflict theory" is not cast in a formal theoretical framework, however. The authors' claim that it "offers a general theory of decision making, not a theory of choice behavior. It is concerned with how human beings arrive at the key consequential choices of living and working, but not with predicting the actual choices they make" (p. 342). Janis, Defares, and Grossman (1983) discuss the hypervigilance coping pattern in further detail.

Although this general theory commands attention because of its clarity of expression and its appeal to our normal experience, its utility for judgment and decision researchers is limited and apparently intentionally so (see above remarks regarding "choice behavior"). Moreover, there is no direct appeal in this or later work to the contemporary literature of judgment and decision making.

Mandler (1982)

Although Mandler would perhaps better be described as an experimental psychologist than as a clinical/personality/social psychologist I include his approach in Literature I because of its broad intent. Mandler offers a theory of stress that is labeled "interruption theory." "The basic premise of interruption theory is that automatic activity results whenever some organized action or thought process is interrupted. . . . That is, any event, external or internal to the individual, that prevents completion of some action, thought sequences, plan, or processing structure is considered to be interrupting. . . . It is important to note that interruption should not be imbued with negative characteristics; this process simply and neutrally involves the disconfirmation of an expectancy or the noncompletion of some initiated action. Interruption is not synonymous with frustration or other related terms. Interruption may be interpreted emotionally in any number of ways, ranging from most joyful to most noxious" (p. 92).

Mandler relates interruption theory to other theoretical concept of stress, as well as to various aspects of cognition, such as memory, consciousness, and problem solving. He also discusses the problems that extensive previous use of the term "arousal" has caused and suggests a more precisely defined alternative.

Despite Mandler's careful development of "interruption theory," it has had little or no impact on the field of judgment and decision theory, nor is it likely to unless it is brought to bear on contemporary approaches in that field.

Coyne and Lazarus (1980)

Coyne and Lazarus describe a "transactional" model of stress (referred to by Lazarus, 1990, in the quotation above). This model "is explicitly cognitive-phenomenological, emphasizing how the person appraises what is being experienced and uses this information in coping to shape the course of events. . . . The effects of the coping are in turn appraised and reacted to as part of the continuous flow of psychological, social, and physiological processes and events. Stressful commerce with the environment thus involves extensive psychological mediation and reciprocal feedback loops, [which] . . . therefore requires that any comprehensive model of it be developed within a transactional, process-oriented perspective" (p. 145). In addition to describing their model in some detail, Coyne and Lazarus advocate naturalistic studies of stress, at the same time urging researchers to remain aware of laboratory studies. Coyne and Lazarus claim that the transactional model is a "radical redirection" from most of the current concept of stress.

The theories offered by Coyne and Lazarus, Janis and Mann, Mandler, and their coworkers are complex and certainly relevant to the present topic. They remain very general and complex, however, which makes it difficult to generate specific predictions for specific circumstances. Much is dependent upon the subject's definition ("appraisal") of circumstances, and much is dependent upon the theorists' method of appraisal of a given subject's state at a specific time.

Levi and Tetlock (1980)

These authors are general with respect to process but specific with respect to content. They start with the premise that "previous studies have found that the cognitive performance of government decision-makers declines in crises that result in war. This decline has been attributed to crisis-produced stress which leads to simplification of information processing. The present study tested the disruptive stress hypothesis in the context of Japan's decision for war in 1941. Two content analysis techniques . . . were used to analyze the translated records of statements by key Japanese policy-makers. Comparisons between statements made in the early and late periods of the 1941 crisis yielded only weak evidence of cognitive simplification. Interestingly, however, the social context in which statements were made significantly affected the complexity of cognitive performance: Statements made in Liaison conferences (in which policies were formulated) were significantly less complex than statements made in Imperial conferences (in which policies were presented to the Emperor for approval). Theoretical and methodological implications of the results were discussed" (p. 195).

This study, although retrospective, is unique in that it explores the effects of stress on the political decision making by Japanese officials as they consider steps toward war. Thus, the authors chose an approach that favors the complexity of decision making outside laboratory conditions over the rigor afforded by them. Although the theorizing is focused on cognition, the context is more nearly that of political science than psychology, and there is little here that will influence theories of the effect of stress on decision making.

Brecke (1982)

Brecke, a former fighter pilot, presents a theoretical model that "unites the variables of cognitive complexity, time availability, uncertainty, and stress into one coherent model. The model is used to examine current aircrew training and to develop new training strategies for improving judgment performance" (p. 951). The model assumes that (1) "judgment task difficulty can be seen as the resultant vector of cognitive complexity, uncertainty, and the inverse of time availability" and (2) "stress will affect judgment performance in a non-linear fashion: positively up to an individual maximum and negatively beyond that. The stress in a situation requiring judgment can be thought of as consisting of three components: the null-level stress, stress resulting from the difficulty of the judgment task itself, and stress resulting from the interaction of the flight problem and background problem" (p. 954). Brecke describes the lack of training aircrews are given in making judgments in stressful situations. The need is particularly acute in the armed forces because of the extreme combination of variables such as cognitive complexity, uncertainty, time pressure, and stress. Brecke also describes a way to train individuals for difficult judgment situations.

Although Brecke's (1982) article offers a number of hypotheses about decision making under stress, he acknowledges that his suggestions for training people to make decisions under stress have "not [been] tested by either experiment or experience" (p. 957). Thus, Brecke's ideas are interesting mainly because they are derived from his military experience. (Note remarks by Commander Rinn above.)

Reviews

I provide only brief comments on the following reviews because the material covered is generally much broader than the present topic and because it overlaps with that included in the theoretical articles cited above. The articles are presented in chronological order inasmuch as they purport to review prior research.

Coyne and Lazarus (1980) review over 40 articles, but few are cognitively oriented. Mandler (1982) reviews over 40 theoretical and empirical articles on stress and thought processes, but few refer directly to judgment and decision making. Janis et al. (1983) review over 40 articles related to stress and decision making, few of which are cognitive in nature.

Saegert and Winkel (1990), in their review of "Environmental Psychology," note a more sophisticated and modern approach taken by Evans and colleagues, thus:

> Evans & Cohen (1987) go beyond the frequently used classification of stressors into cataclysmic events, stressful life events, daily hassles, and ambient stressors (Baum et al. 1982; Campbell 1983; Lazarus & Cohen 1977) to outline eight dimensions along which environmental stressors vary: perceptual salience; type of adjustment required; value or valence of the event; degrees of controllability; predictability; necessity and importance; duration; and periodicity. They note that the physical nature of environmental stressors has been neglected in favor of psychological and sociological investigations of personal, organizational, and societal factors that influence stress and coping. (p. 448)

2. Literature II: Human Factors

Theories

Conceptual work relating stress and cognition within Literature II is perhaps better characterized in terms of "approaches" rather than theories, with the exception of recent work by Hancock (Hancock & Chignell, 1987; Hancock & Warm, 1989), whose work we consider first.

Hancock and Chignell (1987)

In his chapter with Chignell, Hancock offers some observations before developing a theory of "Adaptive Control in Human-Machine Systems." For example, "it is the interaction between the factors of system complexity and operational magnitude that is driving contemporary technology beyond the unaided control capacity of the human operator" (p. 306); "contemporary systems have begun to emphasize knowledge-based operations where, in addition to consultative interaction with the machine, the human [being] is employed for capabilities such as pattern-recognition and inferential reasoning" (p. 307). Further, the goal of the operational unit "cannot be achieved by considering only *static* characteristics of the interface, but requires instead a *dynamic* and therefore adaptive interdependence" (p. 307), which is followed by: "It is our contention that adaptation is a costly process and is becoming an intolerable burden upon the loaded operator" (p. 308). These pressures lead Hancock and Chignell to develop their approach to stress, thus:

> We offer a new view of stress, which is consistent with some elements of the foregoing arguments. However, our proposal contains a number of unique components which differentiate it from those previously discussed. This position has been generated (Hancock & Chignell, 1985) and elaborated (Hancock, 1986; Hancock & Rosenberg, 1987) in a number of recent reports. Our purpose at this juncture is to provide sufficient information to allow the reader to follow our subsequent argument for mental workload as a form of cognitive stress response, and further to follow how such information provides a vital signal for input to an adaptive human-machine interface.
>
> In our approach, a trinity of stress is represented in three descriptive loci. The first of these is an *input* locus composed of the deterministic physical characteristics, or *signature,* of the dynamic environmental display. The second locus is that of *adaptation* undertaken by the responsive individual to compensate for the perturbations introduced by the input stress. The final locus is an *output* value that represents the efficiency of performance upon an on-going, goal-directed task. As adaptation may be one of these latter goals, and as a task may be regarded as an input stress, it is clear that these three loci may overlap. (p. 312)

Hancock and Warm (1989)

The Hancock and Warm contribution is included here because it is the most recent comprehensive review and theoretical treatment. The authors examine past work on the "effects of stress on sustained attention" and then go on to present "a dynamic model . . . that addresses the effects of stress on vigilance and, potentially, a wide variety of attention-demanding performance tasks" (p. 519). Their conclusions regarding the current 1989 state of theoretical work within Literature II are arresting: "There has been a collective failure of theories that seek to explain vigilance performance. . . . This failure is also true for theories of stress in general, which with few exceptions have exhibited similar stagnation" (p. 524). They further note that "the only theoretical construct that spans the two areas [vigilance and stress in general] is the concept of behavioral arousal" (p. 524). But Hancock and Warm then quote Koelega, Brinkman, and Bergman (1986), thus: "But arousal theory can explain any results, post hoc, and lacks predictive power. The position on the inverted-U curve can only be specified after the experiment, so arousal theory, in its present form, is not amenable to rigorous experimental testing" (p. 525) (see also Hockey, 1986a, pp. 44.37–44.38 for a detailed critique and rejection of arousal theory.)

The general disappointment with theoretical progress in Literature II expressed by Hancock and Warm (1989) deserves to be taken seriously. It is based on a broad, expert understanding of Literature II, particularly as it relates to attention, which is the main cognitive area of interest in this literature. Their disappointment parallels that generally expressed in Literature I (see above quotations from Lazarus's 1990 review article in which, for example, he asserts that "stress measurement has almost never been truly theory driven"). Although disappointment in theoretical development is obvious, pessimism is not. And, indeed, were the participants in these two literatures willing to read (not merely note) the others' literature they might well find a certain convergence that would give rise to encouragement. For apparently both sets of theorists now believe that the concept of stress is too broad and must be far more differentiated than it is at present not only with respect to categories of sources—as we shall indicate below—but also in terms of behavior.

Hamilton (1982)

Hamilton's approach brings us closer to experimentally oriented research on cognition. He favors making a "distinction among types of stress, particularly between stress as an effect and stress as an agent" (p. 105), and he argues "in support of an information processing concept of stress as an *agent* (italics added), where stress as an effect is seen as the consequence of the type and amount of information processing mediated by stressors, which contain and generate stressful information" (p. 105). Hamilton also distinguishes among physiological, cognitive, and psychogenic stressors. His main point concerns overloading short-term or working memory. Because all information used to guide behavior resides in working memory, it follows that stress can overload working memory's limited capacity. Thus, "by definition, cognitive stressors are those cognitive events, processes, or operations that exceed a subjective and individualized level of average processing capacity" (p. 109). This overload can result from a person's experience or inexperience with particular stimuli. Note, however, that "an event does not become a stressor until a cognitive processing system has identified it as such on the basis of existing long-term memory data" (p. 117), a statement that brings him close to Mandler (see above) and plunges him into subjectivity.

Nevertheless, Hamilton's focus on information processing and memory is apt to make his theoretical work a source of interest for researchers in stress and judgment and decision making.

Hockey (1979)

Hockey is even more specific in his theory and research than Hamilton (above); thus, "the aims of this chapter are twofold; firstly, to attempt an integrated survey of research findings in the area of stress and performance and, secondly, to propose alternative methodological and theoretical approaches to the experimental study of stress effects in cognition. In reviewing the literature I have concentrated on two main areas of skilled performance, sustained attention and memory. This is primarily because most work has been done in these two fields and the findings are therefore more reliable. In addition, however, and this may be no accident, these two components may be considered as, in some ways, primary in the organization of skilled behavior" (pp. 141–142). Thus, although Hockey's 1979 work does not focus directly on judgment and decision making, it is certainly relevant; his views are informed and broadly based.

In addition to pointing out the problems caused by referring to stress as both cause and effect, Hockey emphasizes the "widespread and largely uncritical acceptance of the Yerkes-Dodson law in human stress research. I do not want to object to its failure to describe the effects of stress adequately, but it blinds us to the recognition of more fundamental changes in functioning" (p. 144). More important questions are " 'What changes underlie the observations embodied in the Yerkes-Dodson law?' 'Why are high levels of arousal bad for performance?' 'What makes a task difficult?' In general these questions have been side-stepped in favour of circular reasoning and naive operational definitions" (p. 144) (see also Hancock and Warm, 1989).

Hockey makes two recommendations: "Adopt an approach of examining the detailed effects of a single stressor across a range of tasks" (p. 170), develop "a realistic functional model of cognitive behaviour . . . with a closer link with the mainstream theory" (p. 170).

Hockey (1983b)

This is an important anthology despite its 1983 publication date. The authors of the 13 chapters are all experts in the various subdomain they review and comment on. Of particular relevance are the chapters by Hockey and Hamilton on "The Cognitive Patterning of Stress States" and Hockey's chapter on "Current Issues and New Directions." Both chapters are informative in a negative sense, that is, they ignore judgment and decision making research.

In his summary chapter entitled "Current Issues and New Directions," Hockey (1983a) identifies four major themes of the research in Literature II, namely:

1. *The use of arousal theory.* "Unfortunately, it is now clear that arousal is a far more complex process than originally conceived. If we are to continue to attempt to relate bodily and mental function . . . , it is clear that we need concepts more realistically suited to the task" (p. 364).

2. *The recognition of the importance of task demands.* "Clearly, there is a great need for the development of a widely-accepted taxonomy for performance functions if this relationship between stress effects and task demands is to be made more generally useful and applicable to a wide range of work conditions" (p. 366).
3. *The appreciation of individual differences.* "In the present state of our knowledge . . . we will need to examine these [individual] characteristics more closely than is possible using temperament or anxiety inventories. . . . A detailed study of individual behaviour under stress . . . may be a more fruitful line of research in the long run" (p. 367).
4. *Interaction of field studies and laboratory experiments.* "The need to consider practical data forces us into developing theories which have a realistic range of application" (p. 368).

The new directions suggested by Hockey include "broad-band methodology, coping strategies, long-term studies, and real-life behaviour" (p. 372).

Hockey's extensive 1986a review in the *Handbook of Perception and Performance* now emphasizes "general patterns of change": "The approach taken here is to examine different states separately across a range of work situations to detect general patterns of change." Further, "it is emphasized throughout . . . that an assessment of stress effects requires information about the overall *pattern* of performance change across different kinds of function" (pp. 44–2). Although Hockey is a prolific researcher and writer, his contributions are more in the form of seeking integration and coherence among the work of others rather than in theoretical efforts.

Cox (1987)

Cox offers general ideas about stress and behavior in the workplace: "This article outlines the developing consensus on the nature of stress. It offers a definition of stress as a psychological state derived from the person's appraisal of their [*sic*] ability to cope with the demands which are made of them. The article then examines the concept of coping and explores its role in stress theory. . . . The article focuses on. . . . [coping as problem solving] and in so doing it describes the nature of rational models of problem solving, considering their utility and application to stress management" (p. 5). Cox's theorizing is general in nature but does not bring the topic of stress into contact with contemporary theories of cognition.

Paterson and Neufeld (1987)

Paterson and Neufeld describe "the situational determinants of the primary appraisal of threat in a specific and systematic manner. Each potential determinant is broken down and the relevant empirical and theoretical literature is reviewed. Eight propositions about the workings of these factors are presented and discussed. Primary attention is given to the factors of event severity, imminence, and probability of occurrence" (p. 404).

Poulton (1976a)

Poulton challenges existing beliefs about the effects of stress, thus: "There are well known rules that heat, noise, and vibration degrade performance. Yet a number of experiments show that all three stresses can reliably *improve* (italics added) performance, especially in tasks requiring speed or vigilance. Many of the results are not widely known, and those that are known may not be believed, whereas fallacious conclusions, which are consistent with the well-known rules, are sometimes accepted without ever checking up on them. In making recommendations for working conditions, the experiments reporting improvements in performance need to be remembered as well as the experiments reporting degradations. The ideal working environment for particular tasks is not necessarily free from all forms of stress. The questions used to obtain subjective assessment of stress do not usually provide categories to indicate that a stress can be beneficial. Subjective assessments do not necessarily mean what the investigator takes them to mean. They may be based upon a well-known rule and thus be consistent across observers. Yet they may indicate that performance has deteriorated when it, in fact, improved. Thus, subjective assessments are not an adequate substitute for measures of performance. Both subjective and objective measures are required in order to give a reasonably complete picture of the effects of stress" (p. 1193). Although Poulton made these observations in 1976, they remain cogent.

Baddeley (1972)

Baddeley was one of the first to examine the effects of dangerous environments on human performance. He reviews "evidence on human performance in dangerous environments" and suggests that "danger reduces efficiency, except in the case of experienced subjects. Perceptual narrowing is shown to be one source of decrement." He further suggests "that danger increases the subject's arousal level which influences performance by producing a narrowing of attention. The nature of the performance decrement and of adaptation to danger are discussed in this context" (p. 537). The hypothesis of perceptual narrowing is discussed by others (see "cognitive narrowing" below).

In 1983 Idzikowski and Baddeley updated this review. They offered

> a general description of how an individual may respond in a dangerous situation. . . . The magnitude of any response will depend on a number

> of factors: (a) the individual's predisposition towards feeling anxious (trait-anxiety) and being aroused (trait-arousal); (b) the individual's assessment of the dangerousness of the situation and his ability to cope with it; and (c) previous exposure. The precise pattern of physiological and biochemical responses will vary from individual to individual unless the situation is perceived as being extreme. In an extreme situation increases in heart-rate, respiration-rate, skin conductance, and muscle tension can be expected, as well as increases in the secretion of catecholamines and various other hormones. Behaviourally, deterioration can be expected in manual dexterity, in sensory-motor tasks such as tracking, and in performance of secondary tasks. It is probable that secondary task performance is reduced before central tasks are affected. (p. 140)

They also noted that their "findings are interpreted within the general arousal framework, which assumes an inverted U-shaped relation between arousal and performance" (p. 141) (but see Hockey, 1983b).

Friedland and Keinan (1982)

So far as can be ascertained, these are the only researchers who have investigated the efficacy of training for stressful conditions (see Rinn above for a description of such training in the U.S. Navy.) They empirically evaluated " 'graduated fidelity training' whereby the trainee is exposed to gradually increasing stressor intensities" and suggested that "it is potentially more effective than high fidelity training." However, they argue, "two conditions are necessary for the realization of this potential effectiveness. First, the trainee must be informed about the upper limit of the stressor intensity which he might encounter in the course of training. In the absence of such information, graduated fidelity training might become highly ineffective. Second, the trainee has to perceive high quality performance as being instrumental for the removal or attenuation of stressors" (p. 41).

This article is grounded in theory and is one of a small number of studies that compare ways of training people to perform a task under stress. It is uncertain to what extent the methods explored and the results obtained may be generalized to other tasks and stressors. It seems doubtful, however, that studies of this type would be convincing to military officers.

Rothstein (1986)

Rothstein chose the lens model (Social Judgment Theory; see Brehmer & Joyce, 1988) conceptual framework as a specific theoretical context for an empirical study of the effect of stress. He concludes that "lens model analyses indicated that cognitive control deteriorated under time pressure while cognitive matching remained unchanged. This effect was limited to complex cue-criterion environments containing curvilinear forms. The results suggest that the time pressured individual tends to be erratic even while implementing correct policy" (p. 83).

Rothstein's (1986) work carries two implications: (1) it is a model for the experimental examination of the effects of stress on specific, quantified parameters of a theory of judgment; thus, it benefits from both the rigor of the laboratory and the complexity of a general theory of judgment; and (2) it disconfirms the cognitive "narrowing" hypothesis by separating the parameter of cognitive control from that of narrowing. Thus, it shows that it is control that is diminished rather than the scope of attention. No other study makes this separation.

Schwartz and Howell (1985)

These authors also chose to examine the effects of stress on the specific theoretical parameters of Social Judgment Theory (see Brehmer & Joyce, 1988). Variation of task parameters, as well as cognitive parameters, were included in the study design. They found that "display formation had a significant effect when time pressure was involved: subjects reached earlier and better terminal decisions under the graphic than the numerical format. . . . The difference reduced to nonsignificance under self-pacing. . . . although significant improvements were obtained by use of a simple aiding device (calculation of worst-case probabilities). Results are generally consistent with Hammond's cognitive [continuum] theory" (p. 433).

This study is unique in employing theoretically specified conditions of information display conditions that predict different types of cognitive activity under stress and their subsequent effects on performance. Predictions were confirmed.

Reviews

For the most extensive recent review, see the *Handbook of Perception and Performance* (Boff, Kaufmann, & Thomas, 1986).

Baddeley (1972) reviews over 25 experimental and theoretical articles and concentrates primarily on studies of dangerous environments such as deep-sea diving. He also draws parallels from research on performance in other dangerous environments. Although the review is almost 20 years old, it remains relevant to contemporary research (see also Idzikowski & Baddeley, 1983).

Poulton (1976a) cites over 80 articles concerned with performance under stress. Studies involving three stressors—heat, noise, and vibration—are discussed. The author argues that generalizable conclusions are difficult to derive from studies of the effects of these stressors. Inconsistencies among studies are highlighted, although there is little discussion of potential theoretical explanations for these inconsistencies. Hockey's (1979) article reviews over 100 articles related to the effects of stress on cognition and behavior. Much of the material reviewed is now dated and restricted to stimulus-response studies, but Hockey makes important arguments concerning the Yerkes-Dodson law and suggestions for future research.

Hamilton (1982) reviews more than 40 articles related to information processing and stress.

Allnutt (1987) refers to approximately 50 articles related to human factors and accidents, with particular attention paid to military aviation accidents. The emphasis is on errors (human or otherwise) rather than stress, although several kinds of errors, contributing to accidents are distinguished, which the author terms "environment-aided errors." Allnutt's discussion of stress and human error includes (1) a criticism of the simplistic nature of research on the effects of stress, despite the complexity of the phenomenon outside the laboratory, (2) the fact that "objective and subjective reactions to stress are often not well correlated" (p. 861); (3) various ways performance may break down under stress, such as narrowing of attention or "reversion" to well-learned behavior patterns.

Allnutt's review is cogent, if somewhat uneven in its coverage. It is perhaps flawed in that it occasionally offers generalizations without empirical support (see the discussion of "cognitive narrowing" below).

Cox (1987) reviews more than 40 primarily theoretical articles about stress in the workplace. The small number of empirical articles reviewed focus primarily on observational data.

Paterson and Neufeld (1987) refer to more than 90 empirical and theoretical articles, but few are directed toward cognitive processes.

Wickens and Flach's (1988) review of research contains only a brief reference to stress: "Human sampling [of information] is affected by high stress, which restricts the number of cues that are sampled. Those few cues which are sampled tend to be those that the pilot perceives to be the most important. Emphasis here is on 'perceives to be.' The pilot's perception of importance will not always reflect the true situation" (p. 118). The authors then cite the well-known example of the pilots who "became preoccupied with an unsafe landing-gear indication and failed to monitor the critical altimeter readings." They conclude that "in general, the sampling behavior of a well-trained operator will approach an optimal strategy . . . [but] limitations will arise due to limited memory and stress." The authors recommend that the design of information displays "should be such that the most important displays are also the most salient, particularly at times of high stress" in order to "ensure that the pilot's perception of importance agrees with the actual state of the world" (p. 118).

Although these conclusions carry significance for the study of stress and decision making, they are not supported in this chapter by references to empirical studies.

Melton (1982) reviews the research on air traffic controllers and air traffic control systems. It contains little substantive information on the effects of stress and decision making, although it does offer a point of view.

Wiener and Nagel's (1988) book contains 19 chapters, most of which contain material indirectly related to stress and decision making. Authors are mainly human factors researchers (see especially Wickens & Flach above.)

The chapter by Hopkins (1988) reviews research results concerning the performance of air traffic controllers. The author's conclusions regarding the stressful nature of the air traffic controller's occupation are surprising (but see also Smith, 1985):

> It was in the occupational health context that the issue of air traffic control as a source of stress on the controller was raised, a notion which more than a decade of extensive work has finally dispelled. . . . Contentions that air traffic control per se necessarily generates symptoms of stress in controllers, and that controllers as a group suffer chronically from stress problems, cannot be sustained. (pp. 654–655)

Hopkins then explains the "flaw in reasoning" that leads to mistaken conclusions about causal relations involving stress. He further declares that "the preoccupation with stress in air traffic control has in retrospect seemed particularly unfortunate, because it has led to the comparative neglect of a greater problem, namely boredom."

3. Literature III: Psychophysiological and Neuroscience

Theories and Reviews

I do not present at this time either theories or reviews (with the exception of Levine, 1990) from the psychophysiological literature for two reasons: (1) It is my judgment that there are no psychophysiological theories available that purport to describe judgment and decision making processes (as currently conceived); (2) I am not prepared to describe or evaluate such theories or reviews if they were available (but see my comments on Levine, 1990, above).

The literature in neuroscience is concentrated mainly on emotion and reason (e.g., Damasio, 1994; Goleman, 1995; LeDoux, 1996). I discuss these works in the text (see chapter 2).

4. Stressors

In this chapter various conditions used as stressors are described, irrespective of the literature in which they appear, although they are largely from Literature II. In each case (1) feasibility of use, (2) effectiveness as a stressor, and (3) reasons for choice of the stressor are indicated.

At least 14 different conditions have been employed to ascertain the effects of stress on judgment processes. Some conditions are intended to be taken at face value as stressors; others rely on the subjects' performance or evaluation of his or her psychological—usually emotional—state.

Sleep Loss

Only Babkoff, Genser, Sing, Thorne, & Hegge (1985a) and Babkoff, Thorne, Sing, Genser, Taube, and Hegge (1985b) have investigated the effects of sleep deprivation on judgment.

Feasibility of use. Sleep loss holds attractiveness for stress research; it is discomforting but essentially painless, leaves no scars, recovery is simple and quick, all subjects are familiar with it and know that it is harmless, and thus do not fear it. It carries direct implications for many important decision making situations. Its principal drawback is that inducing sleep loss requires considerable time and special circumstances.

Effectiveness as a stressor. There is little doubt that sleep loss has direct physiological results on cortical activity which is almost certainly related to cognitive activity that affects judgment and decision making.

Reasons for choice. Feasibility and effectiveness and representativeness of significant political and military content.

Shock (and Threat of)

Electric shock was used in 1974 by Bacon but apparently not used again until 1987 and then only as a threat by Keinan and his colleagues (Keinan, 1987; Keinan & Friedland, 1984; Keinan, Friedland, & Ben-Porath, 1987).

Feasibility of use. The apparatus is simple, but the use of electric shock at a sufficient level to induce stress certainly means inducing pain. Therefore, it is unlikely to be used in university labs in the foreseeable future.

Effectiveness as a stressor. There is little doubt that electric shock and/or fear of it is stressful.

Reasons for choice. See preceding paragraph.

Dangerous Environments

Baddeley has been the foremost investigator (Baddeley, 1972; Idzikowski & Baddeley, 1983); see also Weltman et al. (1971).

Feasibility of use. These circumstances cannot be employed in university laboratories unless (possibly) the subject is deceived. Effects of such environments can generally be ascertained only retrospectively.

Effectiveness as a stressor. There is little doubt about the effectiveness of this stressor; but anecdotes and retrospective studies offer wide varieties of interpretation.

Reasons for choice. High plausibility of effectiveness.

Time Pressure

This is possibly the stressor most widely used in studies close to judgment and decision making. See, for example, Ben Zur and Breznitz (1981); Payne, Johnson, Bettman, and Coupey (1989); Rothstein (1986); Schwartz and Howell (1985); and Zakay and Wooler (1984).

Feasibility of use. This is a simple, painless procedure with no side effects, with complete, rapid recovery, readily quantified and manipulated, and with direct implications for judgment and decision situations outside the laboratory, all of which makes it highly feasible to use, and which no doubt accounts for it being the most frequently used stressor in relation to judgment and decision making.

Effectiveness as a stressor. May vary, depending on other motivating factors; also difficult to separate effect of simple time limitations and stress due to time limitations.

Reasons for choice. Listed under feasibility paragraph above.

Unrepresentative Training

Friedland and Keinan (1982)

Feasibility of use. On the one occasion in which it has been used it appears to have been highly feasible; no pain or side effects are involved.

Effectiveness as a stressor. Because only one study has been conducted, little is known.

Reasons for choice. Situations that are novel and for which subjects are unprepared occur in many important judgment and decision making circumstances outside the laboratory (e.g., operations of ships, planes, power plants, process plants, etc.)

Fatigue

Christensen-Szalanski (1978); Krueger, Armstrong, and Cisco (1985)

Feasibility of use. Doubtful; inducing fatigue is time consuming and may not be approved by human subjects committees.

Effectiveness as a stressor. Uncertain (but see Hockey, 1983b).

Reasons for choice. A commonly observed stressor in work environments, including the military.

Information Processing/Memory Load

Increasingly frequently recognized as a stressor (see especially Hamilton, 1982; Hockey, 1986b).

Feasibility of use. Highly feasible; researchers know how to do this, and human subjects committees will not disapprove because no pain, discomfort, or side effects will occur; memory load can be easily manipulated and quantified.

Effectiveness as a stressor. Uncertain (see Hockey, 1986b, review).

Reasons for choice. See feasibility paragraph above; also will have direct implications for judgment and decision making in work environments.

Threat

A commonly observed stressor in military situations and elsewhere (see, e.g., Janis, 1983; Keinan, 1987; Keinan et al., 1987).

Feasibility of use. Low; human subjects committees are almost certain to object.

Effectiveness as a stressor. Doubtful; much will depend on specific circumstances; can never be taken for granted that stress was in fact induced.

Reasons for choice. Resemblance to judgment and decision making situations outside the laboratory.

Political Crisis

See Levi and Tetlock (1980).

Feasibility of use. Doubtful, can only be used retrospectively.

Effectiveness as a stressor. Uncertain.

Reasons for choice. Motivation to study judgment and decision making in circumstances in which it often occurs with great consequences.

Accident Avoidance

Malaterre, Ferrandez, Fleury, and Lechner (1988)

Feasibility of use. Moderate; while possible to simulate most aspects of the situation in a laboratory (as with drivers' training simulators), simulated situations are still missing certain kinetic and perceptual components. Field simulations (driving courses) are infeasible because of potential risk to the subjects.

Effectiveness as a stressor. Cannot ensure that a simulated situation is stressful. Only one study has been conducted and it did not contain a direct measure of stress. The actual stressor is short lived; measurements would necessarily focus on post-trauma response.

Reasons for choice. Direct implications for judgment and decision making outside the laboratory.

Heat, Noise, Vibration

Hockey (1970, 1983b); Koelega and Brinkman (1986); Koelega, Brinkman, and Bergman (1986); Poulton (1976a, 1976b); Wright (1974)

Feasibility of use. Can produce discomfort or pain; have the benefits of being readily manipulated, quickly induced, easily quantified, and familiar to the subject. Human subjects committee unlikely to allow severe discomfort or pain.

Effectiveness as a stressor. High; have been studied for decades. Have been shown to have *both* incremental and detrimental effects on performance. Almost certain these stressors affect attention capacities in a nontrivial, complex manner psychologically, as well as physiologically.

Reasons for choice. Commonly encountered stressors, therefore externally valid. Obvious applications. High feasibility in laboratory setting.

Heat, Crowding, Confinement

Shanteau and Dino (1993)

Feasibility of use. High; convenient because it is commonly experienced, easily induced and manipulated, and readily quantified. Can potentially produce severe discomfort which would concern human subjects committee.

Effectiveness as a stressor. Can take considerable time to build up stressful effects.

Reasons for choice. High feasibility; direct implications for judgment and decision making situations outside the laboratory.

Exactingness

Extent to which the decision maker is penalized for failing to make appropriate decisions. Hogarth, McKenzie, Gibbs, and Marquis (1991)

Feasibility of use. Easily implemented into a variety of task situations. Has additional benefits of being familiar to subject and readily induced and manipulated.

Effectiveness as a stressor. Dependent on how subject evaluates the situation; exactingness cannot be separated from the task situation in which it is embedded; therefore, it is hard to determine its relative effectiveness; for example, difficult to separate exactingness from punishment.

Reasons for choice. High external validity. Construct validity questionable due to the task and effectiveness interaction described above.

Workload

Smith (1985)

Feasibility of use. Unclear how this can be convincingly induced in the laboratory, although field studies are highly feasible. If successfully induced, the stressor would be familiar to the subjects and readily manipulated.

Effectiveness as a stressor. Very effective stressor but quantifiable measures difficult to obtain.

Reasons for choice. High external validity and intuitive appeal.

Conclusions

Numerous stressors have been employed, both in the laboratory and field. Knowledge regarding their feasibility of use and effectiveness remains uneven and uncertain.

5. Behavioral Consequences (Dependent Variables)

Behavioral consequences are discussed below in terms of their (1) feasibility of measurement, (2) reasons for choice, and (3) conclusions drawn. The articles described are grouped according to psychological variables/functions studies. All three criteria are described insofar as possible in terms of the authors' views.

Before noting the contributions of individual articles under this heading, it should be noted that "attention" has been studied for several decades and there are thousands of publications on this topic. I have cited below only that small fraction of these studies that are more or less closely related to modern approaches to judgment and decision making. An excellent review of this material is provided by Wickens in Hancock's (1987) *Human Factors Psychology* (pp. 29–80).

Wickens's chapter is divided into "Metaphors of Attention," "Selective Attention," "Divided Attention," "Resources, Practice, and Difficulty," "Attention and Human Error," and "Changes in Attentional Function." Wickens's chapter includes over a hundred references; the reader should consult these in addition to the ones annotated here if his or her interests in judgment and decision making lie in its more peripheral aspects. Wickens concludes by stating:

> The topic of human attention has been of interest to psychologists for over a century (Paulen, 1887). While human knowledge has clearly expanded since that time, the basic human constraints and limitations in processing information have remained unchanged. At the same time, the amount of information that humans are being asked to process, integrate, and understand as they interact with today's complex systems is increasing exponentially. Ironically, this fact remains true even as computer automation takes over many of the functions conventionally assigned to humans. This is because a human must now monitor and understand the automating computers and gracefully assume control if and when the automated system fails, as it often does (Rasmussen & Rouse, 1981). As system complexity grows, the number of things that must be monitored grows with it. The better understanding of human attention will not provide all of the answers necessary for coping with

system complexity, but it will certainly offer a good start. (Wickens, 1987, p. 70)

Perception/Attention

Babkoff, Genser, Sing, Thorne, and Hegge (1985a)

Dependent Measure (DM). Lexical decision task.

Feasibility of measurement. Readily used in the laboratory with a variety of stressors. Large supporting literature is available.

Reasons for choice. High feasibility and obvious external validity to situations involving perceptual discrimination.

Conclusion. Stress results in decrement in the ability to discriminate words from nonwords in both visual fields.

Babkoff, Mikulincer, Caspy, Carasso, and Sing (1989)

DM. Search task and pursuit rotor pattern tracing of veridical and mirrored images.

Feasibility of measurement. High.

Reasons for choice. Directly applies to multiple task situations involving time as a stressor.

Conclusion. Stress significantly reduced accuracy. This decrease was exacerbated by circadian rhythms. They also show that sleep loss produces a phase delay of circadian performance accuracy, resulting in a 2–4 hour delay of peak performance.

Babkoff, Thorne, Sing, Genser, Taube, and Hegge (1985b)

DM. Visual search tasks, vigilance discrimination.

Feasibility of measurement. Lends itself well to laboratory study.

Reasons for choice. High feasibility.

Conclusion. Conclusions difficult to draw because measures were collapsed into test batteries and data from individual measures were not discussed.

Bacon (1974)

DM. Dual task: pursuit rotor and auditory signal detection.

Feasibility of measurement. Dual tasks can be stressful in and of themselves so care must be taken to ensure that one task does not overshadow the other.

Reasons for choice. High external validity to situations in which people must perform simultaneous multiple tasks.

Conclusion. Stress affects capacity limits and attentional control processes by narrowing the range of cues processed. This results from a systematic reduction of responses to low-priority aspects of the situation.

Friedland and Keinan (1982)

DM. Visual search task.

Feasibility of measurement. High; there is a large supporting literature. The method can be easily implemented within the laboratory and can be used in conjunction with most stressors.

Reasons for choice. High feasibility and reasonable external validity to other visual search tasks.

Conclusion. Task mastery and the ability to control stress (in this case threat) important for improving performance while under stress. Study showed that graduated unrepresentative training is superior to high unrepresentative training if (1) subject knows the upper limit of the stressor and (2) subject perceives their performance as instrumental in the removal or attenuation of the stressor.

Hockey (1970)

DM. Attentional selectivity within a pursuit tracking and multisource monitoring task.

Feasibility of measurement. High; readily lends itself to laboratory study.

Reasons for choice. High external validity to visual dual task situations. Can be used with most stressors, except stressors involving time.

Conclusion. The primary task (tracking) improved in noise condition. Centrally located signals were detected better than peripheral signals in the presence of greater noise. Authors conclude that there is greater attentional selectivity with arousal: perceptual narrowing.

Keinan (1987)

DM. Analogies test, scanning of alternatives.

Feasibility of measurement. Easily implemented within the laboratory with a wide range of tasks. Has supporting literature from studies done in nonstressful environments.

Reasons for choice. Externally valid for a variety of choice situations. Easily manipulated and implemented.

Conclusion. Subjects exhibited premature closure and nonsystematic scanning under stress.

Keinan and Friedland (1984)

DM. Visual search task.

Feasibility of measurement. Requires proficiency training otherwise the effect of stress with task performance thereby reducing reliability.

Reasons for choice. High external validity particularly to transfer of training.

Conclusion. Results are ambiguous; the authors suggest that while training under stress requires greater time, better transfer to novel situations is anticipated. However, the data here show that subjects trained in nonstress situations achieved higher performance levels in a shorter period of time and that this advantage was not overridden by the introduction of stress (see also Keinan et al., 1987).

Weltman, Smith, and Egstrom (1971)

DM. Central visual acuity and peripheral light detection.

Feasibility of measurement. High; both tasks are easily implemented in the laboratory.

Reasons for choice. Direct applicability particularly to multi-task situations.

Conclusion. Subjects exhibited peripheral narrowing while under stress, while central visual acuity performance remained high, peripheral light detection performance declined. Stress also induced increased heart rates and longer response times.

Conclusion

Feasibility of measurement. The various measures related to perception readily lend themselves to laboratory study. In addition, this area has an extensive supporting literature. Caution should be exercised when combining perceptual measures with manipulations of time because subjects' strategies may change under time pressure. There is also the potential for floor and ceiling effects with increased stress and extensive training, respectively.

Reasons for choice. High feasibility; perceptual measures lend themselves to use with a variety of stressors and are easily quantified both in terms of completion time and accuracy.

Conclusion. The papers discussed here suggest that stressful environments evoke longer task completion times and lower discriminability. One paper also showed that change in response bias was not a factor in the reduced performance levels. However, most of the studies focus on time decrements rather than performance decrements. Several studies discuss critically the Yerkes-Dodson law as a model of performance under stress.

Thinking

Babkoff, Thorne, Sing, Genser, Taube, and Hegge (1985b)

DM. Mental arithmetic and logical reasoning.

Feasibility of measurement. Readily studied within a laboratory setting.

Reasons for choice. Reasonable external validity particularly to classroom situations. Indirectly related to judgment and decision making tasks.

Conclusion. Difficult to draw conclusions because measures were collapsed into test batteries and individual effects were not examined.

Shanteau and Dino (1983)

DM. Ten cognitive measures were used; results are reported for six (memory recall, anagram solution, IQ measure, creativity measure, calibration, and impression formation).

Feasibility of measurement. High; tasks can be easily used in a laboratory.

Reasons for choice. Not mentioned; this study's goal was to survey a variety of tasks.

Conclusion. Noticeable decreases in subjects' puzzle solving and creativity measures.

Conclusion

Feasibility of measurement. Thinking as a construct, like stress, has not been well defined. There is a supportive literature within the areas of intelligence and aptitude testing. These tasks lend themselves to use with various stressors and can be utilized within the laboratory.

Reasons for choice. Reasonable feasibility. There is a direct application to classroom performance and an indirect application to J/DM situations.

Conclusion. Although only two studies are discussed, these studies showed decreasing performance levels with increasing stress.

Judgment and Decision Making (Including Confidence in J/DM)

Ben Zur and Breznitz (1981)

DM. Risky decision; willingness to gamble.

Feasibility of measurement. While readily used in the laboratory, most choice situations do not involve predefined and knowable probabilities, thus a lack of external validity.

Reasons for choice. Easily induced in the laboratory. Subjects' data is readily compared to some "optimal selection" behavior rather than other less quantifiable dependent measures.

Conclusion. Authors conclude that the stress causes subjects to (1) filter out information, thereby processing only a subset of the available information, (2) process information faster by spending less time on a given piece of information, and (3) in particularly high stress conditions, shift strategies.

They also showed that under stress subjects gave proportionally more weight to negative task dimensions and in doing so made less risky decisions.

Böckenholt and Kroeger (1993)

DM. Choice behavior in a binary decision task with multiattribute alternatives. Subjects chose the person they would prefer to date in a computer dating task.

Feasibility of measurement. High.

Reasons for choice. "Binary choices are quite common in daily life" (p. 212). Allows the investigators to look at criterion-dependent choice (CDC) models which makes specific predictions about the various stages in a decision process "related to the acquisition, evaluation, and aggregation of information" (p. 196).

Conclusions. Changes in certain parameters of the CDC model can account for many of the effects of time pressure on decision making. Under time pressure, there was a lowering of the criterion for the decision, including faster information processing and lessening of decision confidence.

Choo (1995)

DM. Part 1: Auditors' judgment performance measured at two levels, overall job-related judgment performance and specific task-related judgment performance in relation to their responses on a mental stress scale. Part 2: Auditors' performance on a judgment case with varying amounts of time pressure (lab setting).

Feasibility of measurement. High.

Reasons for choice. Direct relationship to the subjects' actual work.

Conclusions. Both parts of the findings support Easterbrook's (1959) cue utilization theory, as shown by the inverted U function formed by plotting increasing stress levels and resulting performance.

Christensen-Szalanski (1978)

DM. Judgments and confidence about expected profit situation involving stocks and predicted market trends.

Feasibility of measurement. Moderate; lends itself to laboratory study but generality to nonmonetary situations is questionable.

Reasons for choice. Optimal performance is readily quantified and there is reasonable feasibility.

Conclusion. Data are discussed within the constructs of a utility model. Results indicate that as stress increased so did subjects' confidence ratings.

Edland (1993)

DM. Multiattribute decision task (using students' grades to choose candidates most likely to succeed in a graduate program). The judgments were then related to the use of four decision rules.

Feasibility of measurement. High.

Reasons for choice. A task that frequently occurs and will be familiar to subjects.

Conclusions. Time pressure decreased the use of rules that focused on negative information and gave more importance to positive information. Because the rules that focus on negative information take an extra cognitive step, using that information less under time pressure is seen as a functional coping strategy.

Edland (1994)

DM. Multiattribute decision task (using students' grades to choose candidates most likely to succeed in a graduate program). The judgments were then related to the use of four decision rules.

Feasibility of measurement. High.

Reasons for choice. A task that frequently occurs and will be familiar to subjects.

Conclusions. Different time-limit conditions lead to different choice behavior. Results indicate that the subjects under time pressure make decisions based on the best aspect of the most important attribute. Subjects given more time judge two attributes as most (and about equally) important. However, it is considered that all subjects may actually have the same strategy. If time runs out, then the decision is made on one attribute, but if more time is allowed, the subjects continue to gather information on other attributes.

Gillis (1993)

DM. A multicue judgment task, with the primary index of accuracy being the achievement correlation (between a subject's judgments and the actual values of the criterion). Using baseball teams' statistics from previous years, subjects were asked to predict the number of wins the teams had that season. These predictions could then be prepared with the actual number of wins for those teams.

Feasibility of measurement. High.

Reasons for choice. As well as evaluating the effect of stress on judgmental accuracy, the author was also able to examine another hypothesis about stress and judgment: that stress narrows the focus of attention.

Conclusions. Results lend some support for the hypothesis that stress impairs judgment, but no support for the hypothesis that impairment is due to a narrowing of the focus of attention.

Kaplan, Wanshula, and Zanna (1993)

DM. Subjects judged the likableness of target persons described by a set of two, four, or six traits.

Feasibility of measurement. High.

Reasons for choice. Allows for examining the effects of time pressure on social judgments, specifically judgments of other people, as well as the individual differences of time pressure on the judgments.

Conclusions. Time pressure leads judges to revert to existing stereotypes. Subjects with a high need for structure revert to their biases when under pressure more than subjects with a low need for structure.

Keinan, Friedland, and Ben-Porath (1987)

DM. Multiple-choice analogies test.

Feasibility of measurement. Dubious construct validity: Although the authors label this multiple-choice analogies test as a decision-making task, it seems more reasonable to label it a "thinking task."

Reasons for choice. Allows for multiple tests within a single experimental session.

Conclusion. Even when physical threat rather than time pressure is used, stress results in nonsystematic coverage of decision alternatives (nonsystematic scanning, responding before all alternatives have been considered, i.e., premature closure), and responding without giving each alternative sufficient consideration (temporal narrowing).

Kerstholt (1994)

DM. Decision strategies under risk: Time allocated to decision phases; and behavioral indices: Information requests and actions taken. Subjects acting as the personal trainer for an athlete monitored a graph on a computer screen that informed them continuously of the overall state of the athlete. As the fitness level decreased, the subject had to decide why and take the appropriate action. There were three possible treatable causes for the decline, the fourth possibility was a false alarm. Asking for information involved a cost, waiting too long to take action or taking the wrong action could result in the athlete's complete collapse. Incentives were related to the overall performance, which resulted from the difference between profits obtained by the correct recovery of the athlete and the cost of information requests, incorrect treatments, and complete athlete collapse.

Feasibility of measurement. Low; requires special technology.

Reasons for choice. Dynamic situation: decisions have to be made in real time with feedback from previous decisions.

Conclusions. Results indicate that except for an overall speedup in information processing time pressure did not affect the weighting process of costs of information and risk. Subjects used the same strategy regardless of time pressure. In consequence, under fast changing conditions the subjects took too many risks.

Kerstholt (1995) (Follow-up on the 1994 article)

DM. Same as in 1994 article with the addition of looking at decision making at a lower operational level, reflecting the amount of effort subjects were willing to spend on the decision process. The author also described the action-oriented (immediately applying a treatment) and judgment-oriented (using diagnostic tests to deduce the most likely disorder) strategy with mathematical equations and compared the actual decision strategies to the optimal ones.

Feasibility of measurement. Low; requires special technology.

Reasons for choice. Tests strategy selection in a dynamic situation: decisions have to be made in real time with feedback from previous decisions. Allowed author to compare decisions made in a dynamic situation with those previously found in static situations.

Conclusions. Overall, subjects dominantly chose the judgment-oriented strategy, rather than the action-oriented strategy although the action-oriented one would have been more optimal. At the operational level, subjects seemed to invest less effort with higher false alarm probabilities, and more effort with more time pressure. Subjects adjusted their amount of intervention to the a priori probability of false alarms and not to amount of time pressure, in contrast to the mathematical model. Decision strategy in a dynamic task is less adaptive than is generally found in studies of static tasks.

Kirschenbaum and Arruda (1994)

DM. Range error (absolute difference between a participant's range estimate of the target and the actual range to the target at time of fire), scenario time, and confidence level (submarine task).

Feasibility of measurement. Low; requires special technology.

Reasons for choice. Direct implications for the experts who were tested. "Range is the most important parameter for both weapon firing and collision avoidance" (p. 407).

Conclusions. In a spatial task a comparison was made of verbal uncertainty and graphic uncertainty. The assumption was that the graphic representation would better match the spatial situation and lead to better performance. The graphic representation resulted in superior range estimates, only when the oceanic noise was high and the environmental information was properly modeled. There was no difference in confidence among the groups.

Krueger, Armstrong, and Cisco (1985)

DM. Frequency of aviator crew judgment errors.

Feasibility of measurement. Although judgment and decision making were not formally measured in this experiment, it is highly feasible to design a similar study with the aim of capturing such judgment errors.

Reasons for choice. Judgment and decision making skill critical for aviator performance.

Conclusion. There were occasional instances of judgment error, in which the crew flew off well-practised courses and made incorrect statements about their position relative to intersections and navigational beacons. These errors apparently resulted from navigational miscalculations and the misreading of instruments.

Levi and Tetlock (1980)

DM. Willingness/decision to go to war.

Feasibility of measurement. Authors used integrative complexity coding and cognitive mapping analyses to examine the statements from Japanese policymakers regarding their 1941 decision for war. Although quite complicated, this dependent measure could be used in studies of group decision making within a laboratory setting.

Reasons for choice. Posthoc analysis: Provided a method for studying historical data about a situation that was inherently stressful.

Conclusion. Only weak support was provided for the hypothesis that stress should produce simplified treatment of the decision situation and that integrative complexity of decisions would decrease as stress increased.

Liu and Wickens (1994)

DM. Two mental workload measurements: subjective assessment and time estimation in a decision and monitoring task, involving assigning customers to the shortest of three parallel service lines displayed on a computer monitor. This task was either performed manually by the subject, or automatically by the computer. In both cases, the subjects were also required to monitor for errors in the assignments (their own errors or the computer's errors). Subjective assessment of workload was measured by use of the NASA Weighted Bipolar Workload Rating Scale. This scale uses a weighted combination of six subscale ratings as an integrated measure of mental workload. In order to measure the subjects' time estimation, they were instructed to press a key every 10 seconds during the task.

Also measured were subjects' performance on the task of assigning customers to the correct line and their performance at error detection.

Feasibility of measurement. Low; requires training of assistants.

Reasons for choice. Authors feel that "the evaluation of mental workload is becoming increasingly important in system design and analysis" (p. 1843).

Conclusions. Subjects' performance on the assigning task was significantly faster in the condition with more time pressure than in the condition with less time pressure. Time pressure had no effect on error detection. Both automatic conditions were rated as significantly less workload than the manual conditions. Time pressure alone did not significantly affect subjective workload. There was, however, an interaction between mode (manual or automatic) and time pressure, with the workload rating being significantly higher in the manual condition with higher time pressure. The other three groups were not significantly different from each other, although the automatic with time pressure was rated as a slightly higher workload than the automatic with less time pressure. Time estimation was not affected by time pressure.

Malaterre, Ferrandez, Fleury, and Lechner (1988)

DM. Appropriateness of accident avoidance decisions.

Feasibility of measurement. Data analyzed in terms models alternate choices in accident avoidance situations.

Reasons for choice. Posthoc analysis of accident avoidance performance.

Conclusion. Conclusions tied too closely to accident avoidance to discuss here.

Payne, Bettman, and Johnson (1988)

DM. Willingness to accept risk for financial gain.

Feasibility of measurement. Can be readily used in the laboratory.

Reasons for choice. Method allows for examination of both strategies and strategy selection. Can be generalized to other risky situations, although it is questionable whether outcomes probabilities are as clear in nonlaboratory situations.

Conclusion. The results indicate that when stressed subjects process information more rapidly, increase their selectivity, and tend to use attribute-based processing.

Raby and Wickens (1994)

DM. Student pilot's performance, in a simulator, on three different approaches and landings. Each approach had two main parts. The first was an 11-DME (Distance Measuring Equipment) arc, in which the pilots try to fly an arc of a constant 11-mile radius around the airport. The second part was flying the straight-in landing course. The pilots were evaluated on several performance measures: (1) Five continuous measures of flight performance. Two measures were deviations from the arc and included tracking error and altitude error. The other three were deviations on the final approach course and included airspeed, glideslope, and localizer deviations. (2) The experimenter also electronically recorded the time of initiation and completion of six different discrete tasks. These tasks, which varied in priority, included map reading, answering ATC (Air Traffic Control) communications, setting up instruments and radios, performing safety checks, calling the ATC or asking questions, and completing paperwork.

Feasibility of measurement. Low; requires special technology.

Reasons for choice. The experimenters wanted controlled experimental data from realistic flight simulations to form a systematic understanding of the task management breakdowns that occur in heavy workload periods.

Conclusions. Student pilots sacrificed some aspects of primary flight control as workload increased. Increasing workload increased the amount of time spent performing the high-priority discrete tasks. Increasing workload decreased the time spent on low-priority tasks. Increasing workload did not affect the duration of performance episodes or the optimality of scheduling tasks.

Ramsey, Burford, Beshir, and Jensen (1983)

DM. Unsafe work behaviors.

Feasibility of measurement. This was an extensive and realistic sample of the effects of working conditions that would be difficult to re-create in the laboratory. Other less-extensive studies would be feasible.

Reasons for choice. High generality to stressful working situations.

Conclusion. Temperatures above or below the preferred range of 17–23 degrees Celsius are correlated with a higher number of unsafe work behaviors. This inverse U-shaped function also changes as a function of workload.

Schwartz and Howell (1985)

DM. Optional stopping decision paradigm (hurricane tracking).

Feasibility of measurement. Moderate; the time required to unfold the decision problem could pose problems for experimental use.

Reasons for choice. Highly believable problem situation, in which information becomes available over the course of the problem-solving process. Subjects do not simply select the best alternative rather, they must decide to act or to seek more information.

Conclusion. When stressed the graphic display of information resulted in significantly better performance than the numerical display. Also, performance improved when subjects were aided in calculating worst-case probabilities.

Serfaty, Entin, and Volpe (1993)

DM. Four-member Naval CIC (Combat Information Center) team makes identifications on air targets.

Feasibility of measurement. Low; requires special technology.

Reasons for choice. Direct implications for the subjects, who were in a research course in the Command, Control, and Communications curriculum at the Naval Postgraduate School.

Conclusions. Increasing target uncertainty (in this case, level of jamming) did not have a direct effect on team error rate for target identification. An increase in ambiguity, caused by the similarity of hostile and neutral targets, caused an increase in the rate of team errors. Under high time pressure, teams were able to maintain a very low error rate, not significantly different from that under low time pressure. However, under moderate time pressure, there were significantly more errors made. Under high time pressure, there was a switch in team coordination from an explicit mode to an implicit mode. This change was not seen in the moderate condition, and the switch to an implicit mode was not seen.

Shanteau and Dino (1983)

DM. Several paper and pencil tasks.

Feasibility of measurement: Easily implemented and used with various stressors.

Reasons for choice. High external validity and high feasibility.

Conclusion. Found no effects of stress on complex decision making task.

Shanteau and Dino (1993)

DM. A series of simple (attention and memory) and complex tasks (problem solving and decision making). Specifically, these tasks included: (1) memory and recall tasks consisting of immediate and delayed recall tasks; (2) anagram solutions; (3) IQ measures; (4) creativity tasks consisting of tests designed to measure the quantity and quality of ideas people generate in work situations; (5) calibration tasks to evaluate the subjects ability to accurately assess the state of their own uncertainty; (6) impression formation task, in which subjects are asked to form an impression of a person described by several personality or character traits.

Feasibility of measurement. High.

Reasons for choice. Chosen to evaluate the cognitive ability of subjects in a comfortable, nonstressful situation and again in an uncomfortable, noisy, and crowded situation, to test the effects of environmental stressors.

Conclusions. Creativity was reduced in the stressful environment, although the length of time in the stressful environment did not further affect performance. No stressor effects were found in any of the other tested areas including decision making.

Stiensmeier-Pelster and Schürmann (1993)

DM. Experiment 1: Students had to decide which of 36 dice games they wished to play. Possible reward and probability of winning varied for each game. Subjects were told that each chosen game had to be played after they made their choices on all 36 games. Experiment 2: Students of economics, imagining that they were stock brokers interested in estimating their expected stock exchange profits, were given the use of one out of six packages of information ranging from (1) no information at all to (6) all the information needed to make the correct estimation. If they ignored part or all of the available information, subjects could save time, but this led to inaccurate estimations. The gain in information decreased with each additional new item. Experiment 3: Subjects worked on sets of four risky gambles. Each gamble in a set offered a possible outcome and a probability for the outcome.

Feasibility of measurement. High.

Reasons for choice. Allowed authors to examine time pressure and action versus state orientation (Experiment 1), a more direct assessment of the information processed and the time required (Experiment 2), and a better assessment of acceleration and filtration and to analyze the decision-making process of action- and state-oriented individuals in more detail (Experiment 3).

Conclusions. The action-oriented respond to time pressure by filtering the available information. The state-oriented accelerate their information processing.

Svenson and Benson (1993a)

DM. Senior high school students were tested on decision problems, with two alternative answers.

Feasibility of measurement. High.

Reasons for choice. Easy to use for evaluating time constraints on the framing of decisions. (Framing refers to the different wording of the same problem so that the answers emphasize either the gains or the losses. It has been shown that this change in wording can reverse the preference for the alternatives.)

Conclusions. Time pressure reduced framing effects in the problems that included alternatives that were associated with relatively big losses.

Svenson and Benson (1993b)

DM. Multiattribute decision task (using students' grades to choose candidates most likely to succeed in a graduate program).

Feasibility of measurement. High.

Reasons for choice. A task that frequently occurs and will be familiar to subjects.

Conclusions. This experiment showed that genuine time pressure effects cannot be induced simply through an instruction that there is a scarcity of time. Instead, subjects told that they would have less time than is really needed for the task, but given an adequate amount of time, used the strategy usually used by subjects with plenty of time. Surprisingly, subjects that were told they had ample time, and given the adequate amount of time, acted as if they were under time pressure.

Verplanken (1993)

(In addition to time pressure, need for cognition (NC) was used as an independent measure).

DM. Information-acquisition and decision-making task (billed as a consumer decision task) using an information display board and measuring variability across alternatives and pattern.

Feasibility of measurement. High.

Reasons for choice. Specifically interested in examining individual differences in information search and decision making processes.

Conclusions. Time pressure caused the subjects to speed up their processing and to report less confidence in their decisions. Time pressure resulted in differences in search strategy only among subjects low in need (low-NC) for cognition and not for subjects high in need for cognition (High-NC). Low-NC subjects use more heuristic strategies as indicated by their search strategies being more variable in amount of information across alternatives.

Wallsten (1993)

DM. Computer-controlled, probabilistic decision task. Subjects were to decide whether a sample of information with five binary dimensions more likely came from population A or population B.

Feasibility of measurement. High.

Reasons for choice. Allows experimenters to analyze which dimensions are used in different conditions. Specifically, do subjects in a time pressure situation use a salient subset of the dimensions rather than all of the dimensions, and do subjects in a time pressure situation with a high payoff function attempt to process more dimensions than those with a moderate payoff function?

Conclusions. Time pressure causes subjects to attend to dimensions in decreasing order of salience.

Wickens, Stokes, Barnett, and Hyman (1993)

DM. Pilots' performance on seven variables. Four of the variables relate to response selection: decision choice, optimality, decision time, and decision confidence.

Feasibility of measurement. Low; special technology required.

Reasons for choice. Faulty pilot judgment is considered the cause of most aircraft accidents. Experiment allows real-time analysis of pilots' judgments and decision making.

Conclusions. Performance on the task was significantly degraded by the induced stress. Specifically, performance was degraded for problems with high spatial demands but not for those with high demands on verbal working memory or long-term memory.

Wright (1974)

DM. Consumer choice (car purchase).

Feasibility of measurement. High; however, the information display method could potentially force DM to use unnatural decision strategies.

Reasons for choice. High feasibility.

Conclusion. Stress caused subjects to pay more attention to negative choice attributes. Also, subjects showed dimensional selectivity: They attended to fewer relevant dimensions when making their decisions.

Zakay (1985)

DM. Nursing care decisions.

Feasibility of measurement. High; while the scenarios were artificial, the testing method is a reasonable test of nurses' judgment behaviors.

Reasons for choice. Highly important to understand the situations that produce shifts in decision strategies.

Conclusion. Stress resulted in greater reliance on noncompensatory decision strategies: Important attributes are given proportionally more weight on the final decision.

Zakay and Wooler (1984)

DM. (See conclusion)

Feasibility of measurement. High; however, method might induce the use of untypical strategies.

Reasons for choice. Good feasibility and reasonable external validity to other single choice selection tasks.

Conclusion. Training is ineffective if decision is made under time pressure: Time pressure negates the positive effects of training. However, it is not clear how effective it was to begin with because of (1) the allotted training time was short, and (2) subjects were not allowed to practice.

Conclusion

Stress has been shown to affect (1) the amount of attention and time allotted to the decision task; (2) the thorough study of decision alternatives and issues affecting the selection of alternatives (dimension filtering); (3) the relative weight the decision maker places on negative versus positive expected outcomes of a given choice; and (4) the choice of strategy applied to the task.

Memory

Babkoff, Thorne, Sing, Genser, Taube, and Hegge (1985b)

DM. Digit recall task, serial addition and subtraction.
Feasibility of measurement. High.
Reasons for choice. Reasonable external validity, particularly to classroom situations.
Conclusion. Because the various measures were analyzed as test batteries, it is difficult to describe the specific effect of stress on any one measure.

Krueger, Armstrong, and Cisco (1985)

DM. 6, 8, or 10 digit alphanumeric recall task.
Feasibility of measurement. High; employed standard memory tasks with a strong supporting literature.
Reasons for choice. To-be-recalled digits were similar to military map grid coordinates: Externally valid to such a task.
Conclusion. Authors did not discuss the results of this measure, however, they stated it was a good indicator of stress.

Shanteau and Dino (1983)

DM. Ten cognitive measures were used; results are reported for six (memory recall, anagram solution, IQ measure, creative measure, calibration, and impression formation).
Feasibility of measurement. Authors used simple list-learning paradigm to study immediate and long-term recall.
Reasons for choice. Memory as seen as an essential component of most decision making tasks.
Conclusion. Stress produced small decrements in performance, particularly in the delayed recall condition. However, no substantial changes in serial-position curve shape were observed.

Conclusion

Feasibility of measurement. High.
Reasons for choice. Obvious connections to judgment and decision making situations.
Conclusion. This area has not been explored in great detail and the results are quite varied. This seems in part to be a function of the wide range of theoretical perspectives used when conducting the original work.

Affect (As It Is Related to Above Cognitive Functions)

Babkoff, Thorne, Sing, Genser, Taube, and Hegge (1985b)
(Physiological Variables; Mood Questionnaire)

DM. Adjective checklist of current feelings, Psychiatric symptoms test.
Feasibility of measurement. High.
Reasons for choice. Interesting to relate stress to affect and affect to decision making.
Conclusion. Paper offers little discussion of the affect measures by themselves so direct conclusions about these measures cannot be drawn.

Nasby (1996)

DM. Decision latencies on three orienting sets, self-reference, friend-reference, and experimenter reference.
Feasibility of measurement. High.
Reasons for choice. Examines affect and judgment.
Conclusion. "Elated mood facilitated positive judgments about the self, best friend, and the experimenter, whereas depressed mood facilitated negative judgments about the self only" (pp. 367–368).

Smith (1985)

DM. Air traffic controller performance.
Feasibility of measurement. Not applicable.
Reasons for choice. High external validity.
Conclusion. Paper reviews numerous studies of stress effects on air traffic controller (ATC) performance. The author's general conclusion is that there is little evidence to suggest that ATC's experience inordinate amounts of stress on the job.

Wells and Matthews (1994)

DM. Choosing and implementing a coping strategy as assessed with the Method and Focus of Coping measure developed by Billings and

Moos (1981). This measure consists of 19 yes/no items asking subjects how they coped with a recent stressful life event. The responses are then grouped into three methods of coping categories: active-cognitive; active-behavioral; and avoidance coping. The same responses are also grouped into two focus of coping subscales: problem-focused and emotion-focused. This subscale was based on the original conceptualization of coping of Folkman and Lazarus (1980). A third type of coping, suppression, (Parkes, 1984) was also included.

Also measured were dispositional self-attention using the Private Self-Consciousness subscale of the Self-Consciousness Scale (PSC: Fenigstein, Scheier, & Buss, 1975), and individual differences in the tendency to experience cognitive failures using the Cognitive Failures Questionnaire (CFQ: Broadbent, Cooper, FitzGerald, & Parkes, 1982).

Feasibility of measurement. High.

Reasons for choice. Not clear.

Conclusions. Significant negative correlations between self-attention and emotion-focused coping only in mixed controllability situations. A significant negative correlation between cognitive failures and suppression was also dependent on the subjects' situation-specific control appraisals.

Conclusion

Feasibility of measurement. Moderate; because affect itself is a construct requiring its own definition, it has been difficult to settle on agreed-upon measures or a research paradigm.

Reasons for choice. Stress is accepted as having a large influence on affect; affect is also assumed to be an indicator of stress.

Conclusion. The few studies discussed here draw diverging conclusions.

Learning

Hogarth, McKenzie, Gibbs, and Marquis (1991)

DM. Multiple cue probability learning.

Feasibility of measurement. Highly feasible.

Reasons for choice. Learning an obvious feature of many occupations.

Conclusion. "Exactingness" has an effect on performance (e.g., "performance is an inverted U-shaped function of exactingness").

Rothstein (1986)

DM. Multiple cue probability learning.

Feasibility of measurement. Highly feasible.

Reasons for choice. Learning an obvious feature of many occupations.

Conclusion. "Time pressured individual tends to be erratic even while implementing correct policy." (Note: This result contradicts the "narrowing" hypothesis.)

Conclusion

Feasibility of measurement. High; learning has been studied in a wide variety of methods.

Reasons for choice. Obvious relevance to many cognitive, behavioral activities.

Conclusion. The two studies reported were conducted within the lens model framework, suggesting its utility for this topic.

6. Results of Empirical Studies (Grouped According to Stressor)

The results of research are grouped in terms of the stressors employed as it is the simplest method. Other choices lead to undue complexity because of overlap of categories, fuzzy boundaries, and/or multiple listings.

Sleep Loss

Babkoff, Genser, Sing, Thorne, and Hegge (1985a)

Results/conclusions. "Response lapses increased as a function of sleep loss and were fitted best by a composite equation with a major linear component and a minor rhythmic component. Response accuracy decreased as a function of sleep loss, with the rate of decrease being greater for nonwords than for words. Although d' was higher for right visual field (RVF), it decreased for both fields almost linearly as a function of sleep deprivation. The rate of decrease for RVF stimulation was greater than for left visual field (LVF) stimulation. [Beta] did not change monotonically as a function of sleep loss, but showed strong circadian rhythmicity, indicating that it was not differentially affected by sleep loss per se" (p. 614).

Comments. This article is one of two (see also Babkoff, Thorne, et al. 1985b) that examines the effects of sleep loss as a stressor on the performance of a specific cognitive task.

Babkoff, Thorne, Sing, Genser, Taube, and Hegge (1985b)

Results/conclusions. "As sleep deprivation continued, the average time on task increased at an accelerating rate. The rate of increase differed among tasks, with longer tasks showing greater absolute and relative increases than shorter ones. Such increases confound sleep deprivation and workload effects. In this article, we compare the advantages and disadvantages of several experimental paradigms; describe details of the present design; and discuss methodological problems associated with separating interactions of sleep deprivation, workload, and circadian variation with performance" (p.604).

Comments. The participants in this study were assessed with a substantial test battery. In this article, however, the authors chose to report only the overall time it took to perform tasks of various kinds and did not report more specific measures of performance.

Shanteau and Dino (1993)

Results/conclusions. Subjects were given a series of tests before entering an environmental chamber. The chamber was climate controlled. Subjects in the chamber were (1) completely confined for 24 or 48 hours. (2) Normal sleep patterns were disrupted with 4-hour alternating work/rest schedules. (3) While awake, subjects were given the series of tests. (4) Subjects also were required to ride an exercycle for two 15-minute periods each 24 hours. (5) Normal eating and hygiene habits were disrupted.

Comments. Stress had a definite negative effect on the creativity measures. Surprisingly, the performance on these measures did not decline more as more time was spent in the chamber or with more heat and humidity. The authors speculate that if creativity is impaired in this sort of stressful situation, then so may be the adaptive behaviors that might be required in dealing with a real emergency.

There were no effects of the stressors on the other measures, including decision making.

Shock

Bacon (1974)

Results/conclusions. "Results indicate that arousal narrows the range of cues processed by systematically reducing responsiveness to those aspects of the situation which initially attract a lesser degree of attentional focus. This stimulus loss under arousal represents, independently of any response criterion changes, an actual diminution in the Ss' sensitivity. In addition, it seems that arousal mediates its effect not so much by impeding the initial sensory impression as by affecting the capacity limitations and attentional control processes operating within short-term memory" (p. 81).

Comments. This study is one of a few that uses Signal Detection Theory (cf. Babkoff et al., 1985a) to distinguish sensitivity from response bias in performance of a task under stress. It may even be unique in briefly discussing the possibility that stress affects short-term memory processes rather than perceptual encoding processes.

Friedland and Keinan (1982)

Results/conclusions. The empirical evaluation of " 'graduated fidelity training' whereby the trainee is exposed to gradually increasing stressor intensities . . . suggested that it is potentially more effective than high fidelity training. However, two conditions are necessary for the realization of this potential effectiveness. First, the trainee must be informed about the upper limit of the stressor intensity which he might encounter in the course of training. In the absence of such information, graduated fidelity training might become highly ineffective. Second, the trainee has to perceive high quality performance as being instrumental for the removal or attenuation of stressors" (p. 41).

Comments. This article is grounded in theory and is one of a small number of studies that compare ways of training people to perform a task under stress. It is uncertain whether the methods explored and the results obtained may be generalized to other tasks and stressors.

Keinan and Friedland (1984)

Results/conclusions. "The results pointed to three conditions for the enhancement of training effectiveness: (1) minimal interference of exposure to stressors with task acquisition, (2) familiarity with stressors characteristic of the criterion situation, and (3) absence of unrealistic expectations about future stressors. However, none of the five training procedures meets all three conditions. Implications for the design of procedures whereby persons can be trained to perform proficiently under stress are discussed" (p. 185).

Comments. This article (see also Friedland & Keinan, 1982) is one of a small number of studies that compare methods of training people to perform a task under stress. The comparison of five different procedures, which yielded ambiguous results, demonstrates the complexity of studying decision making and training in decision making under stress.

Keinan, Friedland, and Ben-Porath (1987)

Results/conclusions. "Stress was found to induce a tendency to offer solutions before all decision alternatives had been considered and to scan such alternatives in a nonsystematic fashion. In addition, patterns of alternatives-scanning were found to be correlated with the quality of solutions to decision problems" (p. 219).

Comments. This article, along with Keinan (1987), chiefly explores the hypothesis that previous studies of decision making under stress using time pressure as the stressor have proven inconclusive due to a potential confound, namely "that a complete, systematic scanning of all available alternatives, and the investment of sufficient time in the evaluation of each, might be physically impossible when time is severely limited" (pp. 221–222). Thus, previous investigators have interpreted their results as being due to stress when, in fact, the results may be attributable purely to time limitations.

Dangerous Environments

Baddeley (1972)

Results/conclusions. "Evidence on human performance in dangerous environments is reviewed and suggests that danger reduces efficiency, except in the case of experienced subjects. Perceptual narrowing is shown to be one source of decrement. It is suggested that danger increases the subject's arousal level which influences performance by producing a narrowing of attention. The nature of the performance decrement and of adaptation to danger are discussed in this context" (p. 537).

Comments. This review of over 25 experimental and theoretical articles concentrates mostly on studies of dangerous environments, such as deep-sea diving, but also draws parallels from research on performance in other dangerous environments.

Idzikowski and Baddeley (1983)

Results/conclusions. A review article. "In this chapter we have considered the effect of fear and danger on performance, subjective state, and bodily reactions. We have concentrated primarily on data provided by experiments conducted in controlled dangerous environments, primarily parachuting and diving, though available evidence from the results of natural disasters and war has also been examined. The evidence suggests that when a situation has induced fear in an individual (as measured by subjective and physiological responses), then a deterioration in the efficiency of performance can be expected, especially in tasks involving sensory-motor skill or divided attention. The findings are interpreted within the general arousal framework, which assumes an inverted-U relationship between arousal and performance" (p. 141).

Weltman, Smith, and Egstrom (1971)

Results/conclusions. "The chamber group showed significantly higher anxiety scores and also a significantly higher heart rate throughout the experiment. There was no difference between the groups with regard to correct Landolt detections, although the chamber group responded somewhat slower. Peripheral detection, however, was severely and significantly degraded in the chamber group. It was concluded that perceptual narrowing had been demonstrated as a result of psychological stress associated with exposure to the 'dangerous' pressure-chamber" (p. 99).

Comments. This article demonstrates the effects of perceived threat on the performance of a task. None of the dependent measures was cognitive in nature.

Time Pressure

Ben Zur and Breznitz (1981)

Results/conclusions. "The results show that subjects are less risky under High as compared to Medium and Low time pressure, risk taking being measured by choices of gambles with lower variance or lower amounts to lose and win. Subjects tended to spend more time observing the negative dimensions (amount to lose and probability of losing), whereas under low time pressure they preferred observing their positive counterparts. Information preference was found to be related to choices. Filtration of information and acceleration of its processing appear to be the strategies of coping with time pressure" (p. 89).

Comments. Ben Zur and Breznitz's results are comparable to those of Wright (1974), who employed a multiattribute decision making paradigm. Based on their results, Ben Zur and Breznitz conclude with an implicit recommendation about the manipulation of time pressure. "Thus, the method of obtaining information about dimensions according to preferences is of greater significance in analyzing information processing prior to decision when the extreme values of the time pressure continuum are investigated" (p. 103).

Böckenholt and Kroeger (1993)

Results/conclusions. Criterion-dependent choice (CDC) models are useful for explaining subjects' acquisition of information in choosing between two multiattribute alternatives. The authors say that "the CDC models posit that decision makers process information about choice alternatives by comparing the alternatives with respect to the subjective evaluation of the features, one attribute at a time. Features of the alternatives are processed attributewise, and the results of the attributewise comparisons are accumulated over the processed attributes. When a person has accumulated enough evidence to be convinced that one alternative is better, the comparison process stops and an alternative is chosen" (p. 197). In this experiment the subjects were to choose, out of two choices, the person they would prefer to date, in a computer dating task.

In a no time pressure (NTP) condition, the amount of information selected by a subject was determined by an additive combination of the two factors of overall attractiveness and attractiveness difference. The subjects considered less data when the overall attractiveness was low and when the attractiveness difference was large. Under time pressure (TP), the effect of both factors was smaller than with NTP. Subjects also focused on the more important attributes under TP. Subjects accelerated their decisions under TP, indicated by decreased time per selected attribute.

The subjects under TP reported lower confidence levels in their decisions than did the NTP subjects.

Choo (1995)

Results/conclusions. In both a field study and a laboratory experiment support was found for Easterbrook's (1959) cue utilization theory. Easterbrook proposed that as stress increased from a low to a moderate amount there is a corresponding reduction in irrelevant cue use, and this reduction initially improves judgment performance. Beyond a certain level of stress, judgment performance decreases because relevant as well as irrelevant cues are ignored. Thus, the graphed relation forms an inverted U.

Comments. Eysenck proposed that a person has a limited working memory capacity and a judgment task competes for the limited capacity. He believed that stress resulted in a reallocation of personal attention and effort from judgment to nonjudgment performance. This reallocation limits the capacity for processing incoming stimuli and leads to poorer performance. The graphed relationship would be a U function. Folkman proposed a coping behaviors theory. There are problem-solving oriented behaviors and emotion-oriented behaviors. The problem-solving behaviors have the goal of reducing stress by such actions as obtaining additional resources and reorganizing time schedules. The emotion-oriented behaviors employ emotional and defensive behaviors such as determination and positive thinking. Together these two coping behaviors govern a person's judgment performance. At relatively low levels of stress, the two behaviors occur with fairly equal frequency, and positive problem-solving behaviors (e.g., seeking help, delegating authority, etc.) and positive emotion-oriented behaviors (e.g., mental alertness and positive thinking, etc.) are used. At moderate levels of stress, it is predicted that problem-solving oriented coping behaviors become dominant, however, the two positive coping behaviors increase their intensity and continue to improve judgment performance. In a high stress situation, emotion-oriented behaviors take over. Even though the types of behaviors have switched, it is still assumed that judgment performance gradually increases. This should result in a linear function when performance and stress levels are graphed.

Edland (1993)

Results/conclusions. The use of four different rules was tested on a multiattribute two-choice decision task (using students' grades to choose candidates most likely to succeed in a graduate program). The four rules were: (1) Maximum rule (Max), subjects choose the alternative with the highest possible grade on any of the three attributes; (2) Minimum rule (Min), subjects choose the alternative that does not have the lowest grade on any attribute; (3) Majority of confirming dimensions (MCD), subjects choose the alternative that is superior on two of the three attributes; (4) Sum rule (Sum), subjects choose the alternative that has the highest value when summing all grades across the three attributes.

Time pressure did not, as it was expected to, increase the use of the Max rule. Min and Sum rules were used most in the choice task. There was no evidence of a consistent change in strategy under time pressure in the choice task.

Subjects also judged the difference, using an attractiveness-difference scale, between the two alternatives in each pair, after their choice had been made. An analysis of these scores indicated that with high time pressure there was an increased use of the Max and MCD rule. This shift was less strong than that found previously by other experimenters, in which subjects were given incomplete information. "Maybe time was so short that a general application of Max was not possible, and a modified version of Max focusing just on the most important attribute was adopted" (p. 155). This shift also means an increase toward positive information. "Two things seem to become important under time pressure, the most important attribute and positive information" (p. 155).

The author speculates that the shift away from the Min rule under time pressure could be due to the fact that the Min rule requires higher demand on cognitive capacity. The Min rule takes a two-step process. First, the minimum value must be detected on one of the alternatives, then the other alternative must be chosen.

Edland (1994)

Results/conclusions. "This study has indicated (1) that individuals who make decisions under time pressure judge one single attribute as far more important than the others while individuals who make decisions with no time pressure judge two attributes as most and about equally important, and (2) that differences occur in the use of rules (No time pressure: Min rule/Time pressure: Max rule/Max important rule)" (p. 290).

The author suggests the need for a more complete theoretical framework for better understanding time pressure.

Kaplan, Wanshula, and Zanna (1993)

Results/conclusions. Subjects were to judge the likability of target persons described by sets of two, four, or six traits. Each set was uniformly negative or uniformly positive. Before participating in the judgments, subjects were administered the Personal Need for Structure Scale (Thompson, Naccarato, & Parker, 1989); those in the upper third were considered high need for structure (HNS) subjects, and those in the lower third were considered low need for structure (LNS) subjects (the middle third were not used in the experiment).

Several patterns of results were found. (1) Under time pressure, judgments of the targets were more moderate than under no pressure. That is, under time pressure, subjects' responses to positively described targets were less positive, and responses to negatively described targets were less negative than for subjects with no time pressure. (2) Responses to larger set sizes were more extreme (called the set-size effect), but this effect was attenuated by time pressure. (3) When not under time pressure, the HNS and LNS subjects did not differ in their judgments, but with time pressure, HNS subjects responded less positively to the positive trait sets, and less negatively to the negative sets, than did the LNS subjects. (4) Under time pressure, the set-size effect was attenuated for the HNS subjects as compared to the LNS subjects.

These results can be summarized by two points: time pressure leads judges to revert to existing stereotypes, and there are individual differences in the extent to which stereotypes will be used as heuristics under time pressure.

Kerstholt (1994)

Results/conclusions. "The findings clearly show the specific effects of a dynamic task on decision-making behaviour, especially with respect to the timing of behavioural responses" (p. 103). The results suggested that there was an overall speedup in information processing, but otherwise the weighting process of costs of information and risk on athlete collapse was not affected by time pressure. Kerstholt suggested that the a priori probability of false alarms influenced the timing and that specific values of the costs for information requests versus the costs of treatments influenced the strategic aspects. Incentive scheme had no affect on the timing, but did influence subjects to request a different type of information.

"The findings indicate that people do not optimally react to the time dimension of decision problems" (p. 89).

Kerstholt (1995)

Results/conclusions. With higher a priori probability of false alarms, subjects waited longer to begin treatment. They appeared to invest less effort in the task and were less accurate. This result supports the idea that subjects are risk averse.

With more time pressure subjects processed the information faster and they requested less information. This is similar to the results seen in a static task. The subjects did not vary the time of intervention (time to start treatment) depending on amount of time pressure, as they should have optimally to minimize the costs of information and actions in false-alarm trials. This adaptive variance of time of intervention has been found in static tasks. Subjects did vary the time of intervention, depending on the a priori probability of false alarms.

The action-oriented strategy would have been the best strategy in this task, but the judgment-oriented strategy was predominantly chosen.

"Taken together, the results suggest that strategy selection in dynamic tasks may not be as adaptive as has been observed in static tasks" (p. 198).

Liu and Wickens (1994)

Results/conclusions. Time pressure did not affect the subjective ratings of workload or time estimation. There was, however, an interaction between mode (manual or automatic) and time pressure, with the workload rating being significantly higher in the manual condition with higher time pressure. The authors suggest that their subjects may have reached automaticity in their performance which lead them to be less sensitive to time pressure, especially under the automatic conditions, with lower workloads.

Payne, Bettman, and Johnson (1988)

Results/conclusions. "A computer simulation using the concept of elementary information processes identified heuristic choice strategies that approximate the accuracy of normative procedures while saving substantial effort. However, no single heuristic did well across all task and context conditions. Of particular interest was the finding that under time constraints, several heuristics were more accurate than a truncated normative procedure. Using a process-tracing technique that monitors information acquisition behaviors, two experiments tested how closely the efficient processing patterns for a given problem identified by the simulation correspond to the actual processing behavior exhibited by subjects. People appear highly adaptive in responding to changes in the structure of the available alternatives and to the presence of time pressure. In general, actual behavior corresponded to the general patterns of efficient processing identified by the simulation" (p. 534).

Comments. This article is unique in its use of a computer simulation as an aid for studying strategy selection in decision making. The within-subjects design strengthens the credibility of the conclusions reached and their representativeness of processes occurring outside the laboratory. The conclusion that "people appear highly adaptive" in the "presence of time pressure" is therefore significant.

Rothstein (1986)

Results/conclusions. "Lens model analyses indicated that cognitive control deteriorated under time pressure while cognitive matching remained unchanged. This effect was limited to complex cue-criterion environments containing curvilinear forms. The results suggest that the time pressured individual tends to be erratic even while implementing correct policy" (p. 83).

Comments. This article explores the effects of time pressure on judgment in the context of the lens model (see also Schwartz & Howell, 1985). It is significant because the method employed separates the change in judgment policy from the consistent execution of the policy.

Schwartz and Howell (1985)

Results/conclusions. "Display formation had a significant effect when time pressure was involved: subjects reached earlier and better terminal decisions under the graphic than the numerical format. . . . The difference reduced to nonsignificance under self-pacing . . . although significant improvements were obtained by use of a simple aiding device (calculation of worst-case probabilities). Results are generally consistent with Hammond's cognitive consistency [*sic*] theory" (p. 433).

Comments. This article is one of only a small number that examine decision making under more dynamic task conditions. In addition, this article studies the relation between time pressure, stress, and display format (graphic vs. numeric) on decision making.

Serfaty, Entin, and Volpe (1993)

Results/conclusions. "An increase in the level of external stressor does not necessarily result in a decrease in the performance level of the CIC (Command Information Center) team" (p. 1230). Increasing target uncertainty by garbling the messages did not have an effect on the identification error-rate of the team. The team increased their use of probes and sensor readings to make up for the information lost. However, an increase in ambiguity, that is, the similarity between the hostile and neutral targets, did cause an increase in the number of team errors. The authors point out that this could be due to the team having reached the maximum achievable performance given the ambiguity. In other words, no additional gathering of information could have reduced the errors for this task condition.

Most important, the teams were able to maintain a very low error rate in a high time pressure condition. This error rate did not significantly differ from that in a low time pressure condition. This result is attributed to a change in team coordination strategy. The team moved from an explicit coordination strategy in the low pressure condition to an implicit coordination strategy in the high pressure condition. This does not mean only that the teams reduced the amount of communication; it also involves a reallocation of communication resources. Specifically, it includes a higher proportion of unsolicited information transfers (called "anticipation"), and a different mix and context of messages. This is in part explained by a shared mental model of the situation among the team members, and mental models about the other team members' functions. These mental models allow for team members' anticipatory behavior, or providing information before it was requested.

In the moderate time pressure condition, significantly more errors were made. The implicit or anticipatory behavior was not exhibited in this condition.

Another condition looked at in this study was whether or not the team leader gave frequent updates on his assessment of the target. If the updates were given, the team was more resilient to an increase in ambiguity or time pressure. It is argued that this sharing of the current assessment allows the team members to have a shared model of the situation, allowing for the anticipatory behavior.

Stiensmeier-Pelster and Schürmann (1993)

Results/conclusions. Subjects were tested and then identified as action-oriented or state-oriented. Individuals are considered to be action-oriented when they "focus on a fully developed action structure, that is, when they simultaneously or successively pay attention to (1) their present state, (2) their future state, (3) the discrepancy between present and future states, and (4) at least one action alternative that can reduce the discrepancy. If one or more of these elements are lacking, a person is said to be state-oriented" (p. 244). Action-oriented individuals use an effective operation of action-control processes that facilitates the enactment of their current intentions. State-oriented individuals show impaired action control. Experiment 1 showed how action- and state-oriented people responded to time pressure in two different ways. The action-oriented subjects showed an increased tendency to filtrate the information, that is, use less information under time pressure and focused on probability information, while the state-oriented subjects increased the amount of information processing. Experiment 2 indicated that decision making differed markedly as a function of time pressure. Under high time pressure, subjects asked for less information and required less time to arrive at their estimations. Again action- and state-oriented people responded to time pressure in two different ways. Action-oriented subjects under time pressure decreased the amount of information processed and state-oriented subjects accelerated their speed of processing; thus, they processed the same amount of information but more quickly. Experiment 3 showed no differences for the action-oriented group with or without time pressure, but the state-oriented group processed information at a faster rate under time pressure. Thus, in all three experiments, the action-oriented subjects respond to time pressure by filtrating the available information, while state-oriented subjects accelerate their information processing.

Svenson and Benson (1993a)

Results/conclusions. There was a main effect of time pressure for three of the six problems, and an interaction for time pressure and framing effects. (Framing refers to the different wording of the same problem so that the answers emphasize either the gains or the losses. It has been shown that this change in wording can reverse the preference for the alternatives.) When time pressure was high, there was less or no effect of framing. When there was no time pressure, there was a stronger effect for framing. "In general, a greater degree of differentiation is positively associated with longer processing time, and therefore time pressure would affect more important problems to a greater extent than less important decision problems" (p. 142).

"The result that different formulations of the same decision problem can lead to different preference orders remains a disturbing fact" (p. 143). The authors point out that decision makers may need to be instructed to look at the same task from different perspectives in order to improve their decision making when framing could be a factor. It may not help to get rid of framing biases by just giving the decision maker more time. In fact, more time might make the decisions worse.

Svenson and Benson (1993b)

Results/conclusions. Subjects in two conditions had the same adequate amount of time to make their decisions. However, subjects were led to believe that the time they had to make their decisions was either longer than really needed (sufficient condition) or shorter than really needed (too-short condition), thus "perceived" time pressure was the independent measure. The change in instruction from "sufficient" to "too-short" time caused a change in decision strategy. In the "sufficient" condition, the Max P rule (decision based mostly on the high grade in psychology) was the most applicable. In the "too-short" condition, the most applicable rule was the Min rule (decision based on the alternative not having the lowest grade in any class). The authors say that the "results indicate more time-pressure-type behavior in the sufficient time instruction group" (p. 164). This is shown by the fact that in earlier studies, subjects cope with a time pressure by decreasing the use of the Min rule, because this rule requires that all the information must be scanned and compared in order to make the decision. The Max rule requires only finding the alternative with the highest grade, it does not require switching to the alternative rule and is thus seen as having one less cognitive step. Therefore, decreasing the use of the Min rule is seen as a functional coping method. This coping method did not occur in the too-short time condition but did occur in the "sufficient" condition in this experiment.

Verplanken (1993)

Results/conclusions. "Under time pressure low-NCs ("need for cognition" subjects) show more *variability* in amount of information acquired across alternatives than do low-NCs who worked unpressured. High NC subjects did not behave differently in this respect in the pressured versus unpressured condition. These results suggest that only low-NC subjects respond differently to time-pressured versus unpressured conditions in terms of their decision strategies employed. This finding may lead to the hypothesis that, compared to high-NC individuals, low-NC individuals might be more adaptive to stressful environments" (p. 248).

"There was no time pressure main effect concerning search *pattern* which might indicate that the manipulation of pressure was too weak in this respect. . . . On the other hand, high- and low-NC subjects did show different patterns of search in response to time pressure in that high-Nc subjects showed a marked switch to less attribute-wise search in the second stage of the search process, whereas this was not evident for low-NC subjects. This result might be interpreted as that time-pressured high-NC individuals worked more systematically than did low-NC subjects" (pp. 248–249).

"The findings in this study indicate that individual differences in cognitive motivation are operating in information search and decision-making tasks. This may provide an impetus for researchers in the field of personality as well as decision-making researchers to further explore motivational factors in decision-making processes" (p. 249).

Wallsten (1993)

Results/conclusions. Subjects under no time pressure, used all dimensions equally under moderate and extreme payoff conditions. Subjects with time pressure, primarily used the earlier presented dimensions with later dimensions contributing less and less. However, subjects in the time pressure condition with the extreme payoff rule, used the five dimensions somewhat more equivalently than those in the moderate payoff group.

The author says this reflects a difference in the boundary placement for the two groups under time pressure.

Wickens, Stokes, Barnett, and Hyman (1993)

Results/conclusions. "Pilots were tested in a microcomputer-based simulation of pilot decision tasks known as MIDIS. Subjects viewed a computer display that contained an operating instrument panel and a text window. The text window was used to display a description of various decision 'problems' as they unfold in the course of a realistic flight scenario" (p. 275). Stress was introduced by imposing four variables simultaneously. These four variables consisted of (1) financial risk, (2) increased workload, (3) distracting noise, and (4) time stress.

There was a clear reduction in performance for the stressed group. This reduction was evident in decision optimality and in lower confidence levels. Problems that are more dependent on direct retrieval of stored knowledge are not more affected by stress. There was an effect of stress manipulations on optimality on the dynamic scenarios. A loss of decision confidence resulted from stress in the static (text-based) scenarios, but no reduction was found in decision optimality.

Wright (1974)

Results/conclusions. "Data usage models assuming disproportionately heavy weighting of negative evidence provided best-fits to a significantly higher number of subjects in the high time pressure and moderate distraction [noise] conditions. Subjects also attended to fewer data dimensions in these conditions" (p. 555).

Comments. The main finding of this study, that subjects under time pressure give more weight to negative information, is corroborated by Ben Zur and Breznitz's (1981) results from a risky choice task.

Zakay and Wooler (1984)

Results/conclusions. "It was found that training resulted in more effective decision making only under the 'no time pressure' condition. Under time pressure the training did not improve the quality of decision making at all, and the effectiveness of the decisions was significantly lower than under no time pressure. It was concluded that specific training methods should be designed to help decision makers improve their decisions under time pressure" (p. 273).

Comments. The topic of this article (training people to make decisions under stress) is of critical importance to the military, yet apparently there is little or no research on the topic; but see Keinan (1987) and Keinan and Friedland (1984).

Unrepresentative Training

Friedland and Keinan (1982)

Results/conclusions. The empirical evaluation of " 'graduated fidelity training' whereby the trainee is exposed to gradually increasing stressor intensities . . . suggested that it is potentially more effective than high fidelity training. However, two conditions are necessary for the realization of this potential effectiveness. First, the trainee must be informed about the upper limit of the stressor intensity which he might encounter in the course of training. In the absence of such information, graduated fidelity training might become highly ineffective. Second, the trainee has to perceive high quality performance as being instrumental for the removal or attenuation of stressors" (p. 41).

Comments. This article is grounded in theory and is one of a small number of studies that compare ways of training people to perform a task under stress. It is uncertain whether the methods explored and the results obtained may be generalized to other tasks and stressors.

Fatigue

Christensen-Szalanski (1978)

Results/conclusions. "A manipulation check revealed failure of the task to produce fatigue . . . [although] all participants reported that they felt more mentally fatigued after each 3-hr session than when they began. Thus, an alternative approach to fatigue is afforded by comparing the data from the first half of each of the two session with those of the second half. . . . the results were significant in the predicted direction" (p. 316). That is, participants were significantly less confident in the accuracy of their responses when they reported being more fatigued.

Comments. Although the main concern of this article was not the effect of stress (or fatigue, as manipulated here), participants were less confident in their answers when fatigued.

Hamilton (1982)

Results/conclusions. Hamilton favors making a "distinction among types of stress, particularly between stress as an effect and stress as an agent" (p. 105). He argues "in support of an information processing concept of stress as an agent, where stress as an effect is seen as the consequence of the type and amount of information processing mediated by stressors, which contain and generate stressful information" (p. 105). Hamilton also distinguishes among physiological, cognitive, and psychogenic stressors. His main point about cognitive stressors concerns their effect on overloading short-term or working memory. Because all information used to guide behavior resides in working memory, it follows that stress information can overload working memory's limited capacity. "By definition, cognitive stressors are those cognitive events, processes, or operations that exceed a subjective and individualized level of average processing capacity" (p. 109). This overload can result from a person's experience or inexperience with particular stimuli. Thus, "an event does not become a stressor until a cognitive processing system has identified it as such on the basis of existing long-term memory data" (p. 117).

Comments. This article covers more than 40 articles related to information processing and stress. Hamilton's idea is that stress can overload the limited capacity of working memory and thus degrade cognition and behavior. Hamilton does not cover the facilitation of processing under stress, however, but concentrates on its negative effects.

Note

Fatigue has been investigated in the human factors literature as much or more than any other stressor.

Noise

Hockey (1979)

Results/conclusions. "The aims of this chapter are twofold; firstly, to attempt an integrated survey of research findings in the area of stress and performance and, secondly, to propose alternative methodological and theoretical approaches to the experimental study of stress effects in cognition. In reviewing the literature I have concentrated on two main areas of skilled performance, sustained attention and memory. This is primarily because most work has been done in these two fields and the findings are therefore more reliable. In addition, however, and this may be no accident, these two components may be considered as, in some ways, primary in the organization of skilled behavior" (pp. 141–142).

In addition to pointing out the problems caused by referring to stress as both cause and effect as early as 1970, Hockey was emphasizing the "widespread and largely uncritical acceptance of the Yerkes-Dodson law in human stress research. I do not want to object to its failure to describe the effects of stress adequately, but it blinds us to the recognition of more fundamental changes in functioning" (p. 144). More important questions are " 'What changes underlie the observations embodied in the Yerkes-Dodson law?' 'Why are high levels of arousal bad for performance?' 'What makes a task difficult?' In general these questions have been side-stepped in favour of circular reasoning and naive operational definitions" (p. 144).

Hockey makes two recommendations: "adopt an approach of examining the detailed effects of a single stress across a range of tasks" (p. 170), develop "a realistic functional model of cognitive behaviour . . . [with a] closer link with the mainstream theory" (p. 170).

Comments. This article reviews over one hundred articles related to the effects of stress on cognition and behavior. Much of the material reviewed is now dated and restricted to stimulus response studies, but Hockey makes important arguments concerning the Yerkes-Dodson law and suggestions for future research.

Kirschenbaum and Arruda (1994)

Results/conclusions. A new maximum likelihood estimation algorithm for submarine decision makers was used. This algorithm produces uncertainty estimates that are based on a rule set that uses the same information as previous verbal indicators but are quantitative. The output of this algorithm can be displayed as a 95% uncertainty ellipse (confidence ellipse). This ellipse describes an area of uncertainty about the estimated position of a target. The purpose of this experiment was to compare this graphic representation with the older, verbal representation on the spatial task of range accuracy of submarine officers.

"Under moderately difficult conditions, subjects who had the confidence ellipse available gave range estimates that were significantly more accurate than the range estimates given by those who had only verbal uncertainties" (p. 416). the better decision-making performance shown by the participants with the ellipse may be due to the appropriate matching of a graphical/spatial area of uncertainty with a spatial problem. This matching of the problem and the depiction of the problem resulted in superior decision-making performance (cognitive fit; Vessey, 1991). The authors also say that it may also have allowed subjects to integrally process (Coury, Boulette, & Smith, 1989) the two most important features of the spatial problem used in this experiment: target bearing and target range.

The authors report that the most striking feature of the results is the contrast between the high- and low-noise conditions. the range estimate advantage for the ellipse presentation held only under the difficult conditions of high oceanic noise and correct modeling. The uncertainty ellipse provided no significant advantage when the oceanic noise was low, regardless of modeling accuracy. In fact, there were no differences in time spent in the Ops sum window (containing the ellipse) under low-noise conditions.

The additional information supplied by the ellipse was apparently not needed in the relatively easy task condition of low oceanic noise. Although the basic problem is difficult for novices, it is relatively simple for the expert naval instructors who participated in this experiment (Kirschenbaum, 1992), and the experts were able to use other well-learned processing strategies to produce good decision performance. Thus, the additional information provided by the uncertainty ellipse was effective only when the task was too difficult to be easily solved without it.

Koelega, Brinkman, and Bergman (1986)

Results/conclusions. Although a review, the conclusions are quoted here because they reflect the ambiguity and confusion produced by the research.

> From a literature review of the effects of noise upon visual vigilance performance, Koelega and Brinkman (1986) concluded that one cannot generalize from the existing data, and that many studies were methodologically flawed. Ninety-eight studies published since 1960 were reviewed, and 21 with similar task demands were examined in detail. Diverse results were reported, but most commonly "negative" results; that is, no effect of noise was found.
>
> It is possible that this general lack of a noise effect can be attributed to a common failure of analytical approaches. The measure of performance most often used in vigilance experiments has been the mean percentage of signals detected in each 20-min, 30-min, or 40-min period. It has been observed by some investigators (e.g., Sanders, 1961; Wokoun, 1969) that noise sometimes has profound effects on performance variability that can far overshadow and mask mean differences. More than two decades ago, some authors suggested that averaging over periods of 10 min or more may be too coarse a way to handle the data (Jerison, 1963; McGrath, 1963). In spite of this admonition, only a few investigators have used a more fine-grained analysis of the effects of noise (Fisher, 1972; 1973; Salamé & Wittersheim, 1978; Woodhead, 1964). None of these, however, utilized a simple sensory monitoring experiment.
>
> Investigators have more recently emphasized the desirability of meticulously examining the details of the results (Broadbent, 1976; Jones, 1983, Smith, Jones, & Broadbent, 1981). Goldstein and Dejoy (1980) stated that an overreliance on gross overall measures of performance has impeded progress toward understanding the effects of noise on performance. Furthermore, based on the literature review mentioned previously, it appears that there is as yet no study using a multivariate analysis of the effects of noise in a visual monitoring task, although it has been noted (Jerison, 1977) that one should be wary of bivariate tests in vigilance experiments. Therefore, the present authors decided to employ such an analysis on data collected in a between-subjects design, manipulating both frequency and regularity of noise stimuli occurrence. The hypothesis was that inspection of the microstructure of responding would reveal effects that are absent in gross response indices. (pp. 581–582) . . .
>
> In conclusion, the present study offers some evidence that the usual way of analyzing data from vigilance experiments may veil the effects of

> independent variables such as noise; that expectancy theory cannot explain the effects of noise on vigilance performance; and that sex differences in monitoring performance may be revealed in some studies, but not in others. (p. 591)

Hockey's paper was not cited.

Koelega and Brinkman (1986) had earlier concluded on the basis of their extensive review that "we know nothing about the effects of variable noise on sustained attention" (p. 465).

In view of these remarks on the state of noise as a stressor, we omit further discussion of it.

Wickens, Stokes, Barnett, and Hyman (1993)

See Time Pressure.

Political Crisis

Levi and Tetlock (1980)

Results/conclusions. "Previous studies have found that the cognitive performance of government decision-makers declines in crises that result in war. This decline has been attributed to crisis-produced stress which leads to simplification of information processing. The present study tested the disruptive stress hypothesis in the context of Japan's decision for war in 1941. Two content analysis techniques . . . were used to analyze the translated records of statements by key Japanese policy-makers. Comparisons between statements made in the early and late periods of the 1941 crisis yielded only weak evidence of cognitive simplification. Interestingly, however, the social context in which statements were made significantly affected the complexity of cognitive performance: Statements made in Liaison conferences (in which policies were formulated) were significantly less complex than statements made in Imperial conferences (in which policies were presented to the Emperor for approval). Theoretical and methodological implications of the results were discussed" (p. 195).

Comments. This study, although conducted after events took place, is unique in that it explores the effects of stress on actual political decision makers.

Accident Avoidance

Malaterre, Ferrandez, Fleury, and Lechner (1988)

Results/conclusions. Subjects' estimates of the minimum distance at which they could turn to avoid an obstacle were significantly smaller than their estimates of the distance at which they could brake to avoid an obstacle. This suggests that a lateral movement might be the best accident-avoidance procedure, yet the available literature on the subject shows that people rarely do anything other than brake, and in many cases are not even aware of an alternative course of action.

Comments. Stress was not directly manipulated in these experiments, although it presumably was a part of the experimental situation. The effects of stress per se on the behavior under study are not explored in detail.

Heat

Shanteau and Dino (1983)

Results/conclusions. "Under stress, subjects showed decreases in creativity, lower reliability in decision making, and shifts in serial-position effects. In contrast, stress had little impact on verbal problem solving, general intelligence, or decision complexity" (p. 362).

Comments. This study examined the effects of stress on several problem-solving and decision making tasks. The finding of decreased creativity under stress is noteworthy.

Shanteau and Dino (1993)

See Sleep Loss.

Learning

Hogarth, McKenzie, Gibbs, and Marquis (1991)

> Feedback from the outcomes of decision confounds two kinds of information. One concerns the nature of the underlying decision-making task; the second how well the decision maker has performed in the task. Within the context of learning a repetitive decision-making task, we examine the effects of *exactingness* or the extent to which deviations from optimal decisions are punished. We hypothesize that exactingness has both positive and negative effects on performance and that performance is an inverse-U shaped function of exactingness. In addition to exactingness, we also consider the effects of incentives and argue that these accentuate the positive and negative effects of exactingness. This leads to predicting specific interactions between exactingness and performance. These predictions, as well as several related issues, are examined in a series of five experiments in which exactingness is manipulated by adjusting the coefficient of a squared-error loss function. Results include the findings that (a) performance is an inverted-U shaped function of exactingness, (b) performance is better under incentives when environments are lenient but not when they are exacting, (c) the interaction between exactingness and incentives does not obtain when an incentives function fails to discriminate sharply between good and bad performance, and (d) when the negative effects of exactingness on performance

are eliminated, performance increases with exactingness. We conclude by discussing our work from both theoretical and practical perspectives and make suggestions for further research. (Abstract)

High Pressure Work Situation

Raby and Wickens (1994)

Results/conclusions. Workload was varied by the amount of communication from the ATC (Air Traffic Control) and by the airspeed requirements. More specifically, in the highest workload the pilots were told to intercept the arc when they were closer to it and were also instructed to fly at a higher airspeed, thus giving them less time to prepare for the final approach. There was no direct manipulation of workload on the final approach and landing.

Five different measures were used to evaluate flight performance. On the arc, lateral tracking and altitude errors were recorded. On the final approach and landing, pilots were monitored for airspeed, glideslope, and localizer errors. Lateral tracking errors increased linearly as workload level increased. Increasing workload did not affect altitude deviations. The increased workload on the arc approach caused an increasing amount of error on the measures evaluated in the final approach course, especially the glideslope measure and the airspeed deviations. There was a similar trend for the localizer error, but it did not reach a significant level. Thus, all measures, except altitude error, showed an increase in performance as the workload increased.

The experimenter also electronically recorded the time of initiation and completion of six different discrete tasks. These tasks, which varied in priority, included map reading, answering ATC communications, setting up instruments and radios, performing safety checks, calling the ATC or asking questions, and completing paperwork. The higher the priority of the task the more closely it was performed at the optimal time, and when the tasks were performed, was not affected by workload. When workload increased, higher priority tasks were performed progressively more often and lower priority tasks were performed less often. However, episodes of task performance did not shorten with increasing workload and scheduling optimality was also not altered by workload. The authors add that "the added resources required to figure out when and how to optimally reschedule a task when the workload becomes high will not be available to perform this rescheduling for the very fact that the workload increase has consumed the needed resources. Instead tasks are simply shed . . . or performed at a degraded level" (p. 236).

Lastly, the authors point out that when the workload was high, the pilots spent more time on high-priority discrete tasks, and thus, less time on flight control leading to a greater amount of error on the control task.

Serfaty, Entin, and Volpe (1993)

See Time Pressure.

Smith (1985)

Results/conclusions. We include this article even though it does not directly address the issue of the effects of stress on judgment and decision making because the task of aircraft controller is assumed to involve considerable stress. Contrary to this assumption, Smith concludes that "there is little evidence to support the notion that ATC's are engaged in an unusually stressful occupation. This is not to say that ATC's never encounter stress on the job; however, it does appear that this is the exception rather than the rule. . . . The demands of air traffic work do not appear to place unusual stress on ATC'S; this professional group appears quite capable of handling requirements of the job without distress. The notion that the occupational group is being pressed to the psychological and physiological limit is clearly unjustified" (pp. 106–107).

Wickens, Stokes, Barnett, and Hyman (1993)

See Time Pressure.

Life Stressors

Gillis (1993)

Results/conclusions. Results lend some support for the hypothesis that stress impairs judgment, but no support for the hypothesis that impairment is due to a narrowing of the focus of attention. Specifically, it was found that "certain stressful circumstances may impair judgments but only if these circumstances lead to an individual experiencing distress at the time the subject is required to make such judgments" (p. 1361). Two of the stresses looked at here, life stresses over a six-month period and dysfunctional thinking, were not directly related to the actual time the judgment took place. State anxiety and depression did relate significantly with impaired judgmental performance. No outside stressors were introduced.

Financial Risk

Wickens, Stokes, Barnett, and Hyman (1993)

See Time Pressure.

Emotions

Nasby (1996)

Results/conclusions. Forgas (1995) suggests that, according to the multiprocess theory of mood and social judgment, mood-congruent judgment (MCJ) frequently occurs under conditions that foster elaborative processing. A social judgment resulting from elaborative processing would occur when a judge relates information about a target to preexisting knowledge. MCJ would naturally occur to the extent that mood selectively increased accessibility of mood-congruent knowledge. Investigators also believe that selective elaboration produces mood-congruent encoding (MCE). The mood states cause a priming of mood-congruent cues that include associations, categories, constructs, recollections of specific events, and schemata. Mood-congruent information can then become linked and therefore elaborated. In this experiment an assessment was made of decision latencies of social judgments involving three orienting sets: self-reference, friend-reference, and experimenter reference.

Elated mood facilitated positive judgments about all three referents, but depressed mood facilitated negative judgments about the self only. The depressed mood also produced MCE, but again, only under conditions of self-reference.

Wells and Matthews (1994)

Results/conclusions. Subjects first recalled a recent stressful event or personal crisis, described it, indicated on a 3-point scale how much the event mattered, indicated on a 3-point scale how much control they had over the situation, and then completed the coping subscales and the two personality measures (described in chapter 5).

Choice of coping strategy is correlated with both situational and personality factors and the interaction between them. The strongest situation predictor, perceived importance of event, indicated increased use of problem-focused coping with reduced suppression.

Effects of personality on emotion-focused coping and suppression were confined to mixed-controllability situations, with Private Self-Consciousness (PSC) being associated with less use of all three coping strategies, and cognitive failures negatively related to suppression only. However, the correlation between PSC and suppression seemed to be statistically mediated by cognitive failures. This implies that "PSC may only affect the two more active strategies of emotion and problem-focused coping."

The data gives broad support for the authors' hypothesis that "self-focus is associated with a reduction in active coping in stressful situations and cognitively demanding situations" (p. 291).

7. Conclusions

No defensible consensus, readily justified on empirical grounds, arises from our examination of these articles. No general principle explaining the effect of stress on judgment and decision making is supported by a conclusive set of empirical studies. Although it has been demonstrated that stress impairs, enhances, and has no effect on cognitive activity on some occasions, with some people, no generalization over conditions or people is secure.

Empirical Work

The empirically based articles describe a wide variety of stressors, but time pressure, noise, and electric shock were usually manipulated experimentally. Stressors that appeared in observational studies are heat and emergency operations of various kinds, for example, controlling air traffic or directing a helicopter simulator through difficult maneuvers. Experimental studies far outnumber observational or field studies.

Theoretical/Review Work

The theoretical work examined also failed to yield a consensus on the causes or effects of stress on judgment and decision making behavior. Theories are largely informal; hypotheses are rarely rigorously formulated.

Nor do the review articles aid in establishing coherence among the articles included. Review articles cover work ranging from performance in dangerous environments such as deep-sea diving to critiques of the noise-as-a-stressor literature. Empirical studies are not organized in relation to theories.

Unresolved Issues

1. Does stress cause "narrowing"? This question has received considerable attention, and the answer has generally, but not always been, "Yes" (see Allnutt, 1987; Bacon, 1974; Baddeley, 1972; Hockey, 1970; Keinan, 1987; Keinan et al., 1987; Weltman et al., 1971; Wright, 1974), and its truth seems to be taken for granted. The "narrowing" hypothesis should receive a high priority in future systematic research.

2. The utility of the Yerkes-Dodson (inverted-U) arousal law is increasingly being questioned. Does it directly apply to judgment and decision making under stress? If so, which cognitive processes are affected, and in what way?

3. How is a specific environmental stressor (e.g., time pressure) to be calibrated on a stressor scale? How is subjective stress to be calibrated so that meaningful comparisons can be made between studies? Without such calibrations, it will be difficult to ascertain the conditions under

which environmental stress produces subjective stress, to determine whether these conditions involve high or low stress, and to determine when (or whether) these conditions lead to impaired, enhanced, or unchanged performance. Until such work is done, no definite conclusions can be reached about prior research because it is unclear where to place each study's stress condition on the stressor scale. Similarly, it is difficult to ascertain the location of the subject's response on the stressor scale.

Author Notes

I wish to acknowledge the considerable assistance I have received in the preparation of this Literature Review from Cynthia Lusk, Paula Messamer, Ernest Mross, Mary Luhring, Vicki Schneider, Tracy Stimson, and especially, Doreen Petersen.

The work for this review was supported in part by the Office of Basic Research, Army Research Institute, Contract MDA903-86-C-0142, Work Unit Number 2Q161102B74F. The views, opinions, and findings contained in this review are those of the author and should not be construed as an official Department of the Army position, policy, or decision, unless so designated by other official documentation.

References

Abelson, R. P., & Levi, A. (1985). Decision making and decision theory. In G. Lindzey & E. Aronson (Eds.), *Handbook of Social Psychology* (3rd ed.) (Vol. 1, pp. 231–309). New York: Random House.

Alexander, C. (1998). *The Endurance: Shackleton's Legendary Antarctic Expedition.* New York: Knopf.

Allnutt, M. F. (1987). Human factors in accidents. *British Journal of Anaesthesia, 59,* 856–864.

Anderson, N. H. (1962). Application of an additive model to impression formation. *Science, 138,* 817–818.

Anderson, N. H. (1965). Averaging versus adding as a stimulus-combination rule in impression formation. *Journal of Experimental Psychology, 70,* 394–400.

Anderson, N. H. (1979). Algebraic rules in psychological measurement. *American Scientist, 67,* 555–563.

Anderson, N. H. (1981). *Foundations of Information Integration Theory.* New York: Academic Press.

Anderson, N. H. (1982). *Methods of Information Integration Theory.* New York: Academic Press.

Anderson, N. H. (1996). *A Functional Theory of Cognition.* Mahwah, NJ: Erlbaum.

Appley, M. H., & Turnbull, R. (Eds.). (1986). *Dynamics of stress: Physiological, psychological, and social perspectives.* New York: Plenum.

Babkoff, H., Genser, S. G., Sing, H. C., Thorne, D. R., & Hegge, F. W. (1985a). The effects of progressive sleep loss on a lexical decision task: Response lapses and response accuracy. *Behavior Research Methods, Instruments, & Computers, 17,* 614–622.

Babkoff, H., Mikulincer, M., Caspy, T., Carasso, R. L., & Sing, H. (1989). The implications of sleep loss for circadian performance accuracy. *Work & Stress, 3*, 3–14.

Babkoff, H., Thorne, D. R., Sing, H. C., Genser, S. G., Taube, S. L., & Hegge, F. W. (1985b). Dynamic changes in work/rest duty cycles in a study of sleep deprivation. *Behavior Research Methods, Instruments, & Computers, 17*, 604–613.

Bacon, S. J. (1974). Arousal and the range of cue utilization. *Journal of Experimental Psychology, 102*, 81–87.

Baddeley, A. D. (1972). Selective attention and performance in dangerous environments. *British Journal of Psychology, 63*, 537–546.

Bar-Hillel, M., & Yaari, M. (1993). Judgments of distributive justice. In B. A. Mellers & J. Baron (Eds.), *Psychological perspectives on justice: Theory and applications* (pp. 55–84). New York: Cambridge University Press.

Baum, A., Singer, J. E., & Baum, C. (1982). Stress and the environment. In G. W. Evans (Ed.), *Environmental stress* (pp. 15–44). New York: Cambridge University Press.

Ben Zur, H., & Breznitz, S. J. (1981). The effect of time pressure on risky choice behavior. *Acta Psychologica, 47*, 89–104.

Berlin, I. (1991). *The crooked timber of humanity: Chapters in the history of ideas.* New York: Knopf.

Berlin, I. (1997). *The sense of reality: Studies in ideas and their history.* New York: Farrar, Straus, and Giroux.

Bernieri, F. J., Gillis, J. S., Davis, J. M., & Grahe, J. E. (1996). Dyad rapport and the accuracy of its judgment across situations: A lens model analysis. *Journal of Personality & Social Psychology, 71*, 110–129.

Billings, A. G., & Moos, R. H. (1981). The role of coping responses and social resources in attenuating the stress of life events. *Journal of Behavioral Medicine, 4*, 139–157.

Blakeslee, S. (1996, January 23). Figuring out the brain from its acts of denial. *New York Times*, pp. B5, B7.

Blass, T. (1991). Understanding behavior in the Milgram obedience experiment: The role of personality, situations, and their interactions. *Journal of Personality and Social Psychology, 60*, 398–413.

Böckenholt, U., & Kroeger, K. (1993). The effect of time pressure in multiattribute binary choice tasks. In O. Svenson & A. J. Maule (Eds.), *Time pressure and stress in human judgment and decision making* (pp. 195–214). New York: Plenum.

Boff, K. R., Kaufmann, L., & Thomas, J. P. (Eds.). (1986). *Handbook of perception and performance: Vol. 2. Cognitive processes and performance.* New York: Wiley.

Brecke, F. H. (1982). Instructional design for aircrew judgment training. *Aviation, Space, and Environmental Medicine, 53*, 951–957.

Brehmer, B. (1974). Hypotheses about relations between scaled variables in the learning of probabilistic inference tasks. *Organizational Behavior and Human Performance, 11*, 1–27.

Brehmer, B. (1996). Man as a stabilizer of systems: From static snapshots of judgment processes to dynamic decision making. *Thinking and Reasoning, 2*, 225–238.

Brehmer, B., & Dörner, D. (1993). Experiments with computer-simulated microworlds: Escaping both the narrow straits of the laboratory and the deep blue sea of the field study. *Computers in Human Behavior*, *9*(2–3), 171–184.

Brehmer, B., & Joyce, C. R. B. (Eds.). (1988). *Human judgment: The SJT view.* Amsterdam: Elsevier.

Brehmer, B., & Lindberg, L.-A. (1970). The relation between cue dependency and cue validity in single-cue probability learning with scaled cue and criterion variables. *Organizational Behavior and Human Performance*, *5*, 542–554.

Brennan, W. (1987). Reason, passion, and the progress of the law. *The Record of the Association of the Bar of the City of New York*, *42*, 948–977.

Broadbent, D. E. (1957). A mechanical model for human attention and immediate memory. *Psychological Review*, *64*, 205–215.

Broadbent, D. E. (1958). *Perception and communication.* New York: Pergamon.

Broadbent, D. E. (1971). *Decision and stress.* London: Academic Press.

Broadbent, D. E. (1976). Noise and the details of experiments: A reply to Poulton. *Applied Ergonomics*, *7*, 231–235.

Broadbent, D. E., Cooper, P. F., FitzGerald, P., & Parkes, K. R. (1982). The Cognitive Failures Questionnaire (CFQ) and its correlates. *British Journal of Clinical Psychology*, *21*, 1–16.

Bruner, J. S., Goodnow, J. J., & Austin, G. A. (1956). *A study of thinking.* New York: Wiley.

Brunswik, E. (1934). *Wahrnehmung und gegenstandswelt: Grundlegung einer psychologie vom gegenstand her (Perception and the world of objects: The foundations of a psychology in terms of objects).* Leipzig and Vienna: Deuticke.

Brunswik, E. (1943). Organismic achievement and environmental probability. *Psychological Review*, *50*, 255–272.

Brunswik, E. (1952). The conceptual framework of psychology. In *International encyclopedia of unified science* (Vol. 1, no. 10, pp. 4–102). Chicago: University of Chicago Press.

Brunswik, E. (1955). Representative design and probabilistic theory in a functional psychology. *Psychological Review*, *62*, 193–217.

Brunswik, E. (1956). *Perception and the representative design of psychological experiments* (2nd ed.). Berkeley: University of California Press.

Campbell, J. M. (1983). Ambient stressors. *Environment and Behavior*, *15*, 355–380.

Carley, W. M. (1993, April 26). Mystery in the sky: Jet's near-crash shows 747s may be at risk of autopilot failure. *The Wall Street Journal*, pp. A1, A6.

Carpenter, P. A., Just, M. A., & Shell, P. (1990). What one intelligence test measures: A theoretical account of the processing in the Raven Progressive Matrices Test. *Psychological Review*, *97*, 404–431.

Chasseigne, G., Mullet, E., & Stewart, T. R. (1997). Aging and multiple cue probability learning: The case of inverse relationships. *Acta Psychologica*, *97*, 235–252.

Choo, F. (1995). Auditors' judgment performance under stress: A test of the predicted relationship by three theoretical models. *Journal of Accounting, Auditing & Finance*, *10*, 611–641.

Christensen-Szalanski, J. J. J. (1978). Problem solving strategies: A selection mechanism, some implications, and some data. *Organizational Behavior and Human Performance*, *22*, 307–323.

Committee on Armed Services. (1989). *Iran air flight 655 compensation* (H.A.S.C. No. 100–119). Washington, DC: U.S. Government Printing Office.

Cooksey, R. W. (1996). *Judgment analysis: Theory, methods, and applications.* San Diego: Academic Press.

Cosmides, L., & Tooby, J. (1996). Are humans good intuitive statisticians after all? Rethinking some conclusions from the literature on judgment under uncertainty. *Cognition, 58,* 1–73.

Coury, B. G., Boulette, M. D., & Smith, R. A. (1989). Effect of uncertainty and diagnosticity on classification of multidimensional data with integral and separable displays of system status. *Human Factors, 31,* 551–569.

Cox, T. (1987). Stress, coping and problem solving. *Work & Stress, 1,* 5–14.

Coyne, J. C., & Lazarus, R. S. (1980). Cognitive style, stress perception, and coping. In I. L. Kutash, L. B. Schlesinger, & Associates (Eds.), *Handbook on stress and anxiety.* San Francisco: Jossey-Bass.

Cray, E. (1997). *Chief Justice: A biography of Earl Warren.* New York: Simon & Schuster.

Dalgleish, L. I., & Drew, E. C. (1989). The relationship of child abuse indicators to the assessment of perceived risk and to the court's decision to separate. *Child Abuse and Neglect, 13,* 491–506.

Damasio, A. R. (1994). *Descartes' error: Emotion, reason, and the human brain.* New York: G. P. Putnam.

Driskell, J. E., & Salas, E. (Eds.). (1996). *Stress and human performance.* Mahwah, NJ: Erlbaum.

Earle, B. (in press). Brunswikian contributions to moral psychology. In K. R. Hammond & T. R. Stewart (Eds.), *The essential Brunswik: Beginnings, explications, applications.* New York: Oxford University Press.

Easterbrook, J. A. (1959). The effect of emotion on cue utilization and the organization of behavior. *Psychological Review, 66,* 183–201.

Edland, A. (1993). The effects of time pressure on choices and judgments of candidates to a university program. In O. Svenson & A. J. Maule (Eds.), *Time pressure and stress in human judgment and decision making* (pp. 145–156). New York: Plenum.

Edland, A. (1994). Time pressure and the application of decision rules: Choices and judgements among multiattribute alternatives. *Scandinavian Journal of Psychology, 35,* 281–291.

Edwards, W. (1954). The theory of decision making. *Psychological Bulletin, 51,* 380–417.

Epstein, S. (1994). Integration of the cognitive and the psychodynamic unconscious. *American Psychologist, 49,* 709–724.

Evans, G. W., & Cohen, S. (1987). Environmental stress. In D. Stokols & I. Altman (Eds.), *Handbook of environmental psychology* (Vol. 1, pp. 571–610). New York: Wiley.

Fechner, G. T. (1966). *Elements of psychophysics* (H. E. Adler, Trans.). New York: Holt, Rinehart, & Winston. (Original work published 1860)

Fenigstein, A., Scheier, M. F., & Buss, A. H. (1975). Public and private self-consciousness: Assessment and theory. *Journal of Consulting and Clinical Psychology, 43,* 522–527.

Fisher, S. (1972). A "distraction effect" of noise bursts. *Perception, 1,* 223–236.

Fisher, S. (1973). The "distraction effect" and information processing complexity. *Perception*, *2*, 79–89.

Flanagan, J. C. (1954). The critical incident technique. *Psychological Bulletin*, *51*, 327–358.

Folkman, S., & Lazarus, R. S. (1980). An analysis of coping in a middle-aged community sample. *Journal of Health and Social Behavior*, *21*, 219–239.

Folkman, S., & Lazarus, R. S. (1988). The relationship between coping and emotion. *Social Science and Medicine*, *26*, 309–317.

Forbes, A. R. (1964). An item analysis of the advanced matrices. *British Journal of Educational Psychology*, *34*, 1–14.

Forgas, J. P. (1995). Mood and judgment: The affect infusion model (AIM). *Psychological Bulletin*, *117*, 39–66.

Fraser, C. (1997, August 14). The revenge of Everest. *New York Review of Books*, pp. 59–64.

Friedland, N., & Keinan, G. (1982). Patterns of fidelity between training and criterion situations as determinants of performance in stressful situations. *Journal of Human Stress*, *8*(4), 41–46.

Frisch, D., & Clemen, R. T. (1994). Beyond expected utility: Rethinking behavioral decision research. *Psychological Bulletin*, *116*, 46–54.

Funder, D. (1987). Errors and mistakes: Evaluating the accuracy of social judgment. *Psychological Bulletin*, *101*, 75–90.

Funder, D. C. (1995). On the accuracy of personality judgment: A realistic approach. *Psychological Review*, *102*, 652–670.

Funder, D. C. (1996). Base rates, stereotypes, and judgmental accuracy. *Behavioral and Brain Sciences*, *19*, 22–23.

Funder, D. C., & Sneed, C. D. (1993). Behavioral manifestations of personality: An ecological approach to judgmental accuracy. *Journal of Personality and Social Psychology*, *64*, 479–490.

Gigerenzer, G., & Goldstein, D. G. (1996). Reasoning the fast and frugal way: Models of bounded rationality. *Psychological Review*, *103*, 650–669.

Gillis, J. S. (1993). Effect of life stress and dysphoria on complex judgments. *Psychological Reports*, *72*, 1355–1363.

Gillis, J. S., Bernieri, F. J., & Wooten, E. (1995). The effects of stimulus medium and feedback on the judgment of rapport. *Organizational Behavior and Human Decision Processes*, *63*, 33–45.

Goldstein, J., & Dejoy, D. M. (1980). Behavioral and performance effects of noise: Perspectives for research. In J. V. Tobias, G. Jansen, & W. D. Ward (Eds.), *Proceedings of the Third International Congress on Noise as a Public Health Problem* (pp. 303–321). Rockville, MD: The American SLH Association.

Goleman, D. (1995). *Emotional intelligence*. New York: Bantam Books.

Green, D. M., & Swets, J. A. (1974). *Signal detection theory and psychophysics*. Huntington, NY: Krieger.

Grinker, R. R., & Spiegel, J. P. (1945). *Men under stress*. London: J. & A. Churchill.

Hamilton, V. (1982). Cognition and stress: An information processing model. In L. Goldberger & S. Breznitz (Eds.), *Handbook of stress: Theoretical and clinical aspects* (pp. 105–120). New York: Free Press.

Hammond, K. R. (1948). Subject and object sampling: A note. *Psychological Bulletin*, *45*, 530–533.

Hammond, K. R. (1954). Representative vs. systematic design in clinical psychology. *Psychological Bulletin, 51*(2), 150–159.

Hammond, K. R. (1955). Probabilistic functioning and the clinical method. *Psychological Review, 62*, 255–262.

Hammond, K. R. (Ed.). (1966). *The psychology of Egon Brunswik*. New York: Holt, Rinehart, and Winston.

Hammond, K. R. (1996). *Human judgment and social policy: Irreducible uncertainty, inevitable error, unavoidable injustice*. New York: Oxford University Press.

Hammond, K. R., & Adelman, L. (1976). Science, values, and human judgment. *Science, 194*, 389–396.

Hammond, K. R., Frederick, E., Robillard, N., & Victor, D. (1989). Application of cognitive theory to the student-teacher dialogue. In D. A. Evans & V. L. Patel (Eds.), *Cognitive science in medicine: Biomedical modeling* (pp. 173–210). Cambridge, MA: MIT Press.

Hammond, K. R., & Grassia, J. (1985). The cognitive side of conflict: From theory to resolution of policy disputes. In S. Oskamp (Ed.), *Applied social psychology annual: Vol. 6. International conflict and national public policy issues* (pp. 233–254). Beverly Hills: Sage.

Hammond, K. R., Hamm, R. M., & Grassia, J. (1986). Generalizing over conditions by combining the multitrait-multimethod matrix and the representative design of experiments. *Psychological Bulletin, 100*, 257–269.

Hammond, K. R., Hamm, R. M., Grassia, J., & Pearson, T. (1987). Direct comparison of the efficacy of intuitive and analytical cognition in expert judgment. *IEEE Transactions on Systems, Man, and Cybernetics, 17*, 753–770.

Hammond, K. R., Hamm, R. M., Grassia, J., & Pearson, T. (1997). Direct comparison of the efficacy of intuitive and analytical cognition in expert judgment. In W. M. Goldstein & R. M. Hogarth (Eds.), *Research on judgment and decision making: Currents, connections, and controversies* (pp. 144–180). Cambridge: Cambridge University Press.

Hammond, K. R., Harvey, L. O., Jr., & Hastie, R. (1992). Making better use of scientific knowledge: Separating truth from justice. *Psychological Science, 3*, 80–87.

Hammond, K. R., & Joyce, C. R. B. (Eds.). (1975). *Psychoactive drugs and social judgment: Theory and research*. New York: Wiley.

Hammond, K. R., McClelland, G. H., & Mumpower, J. (1980). *Human judgment and decision making: Theories, methods, and procedures*. New York: Hemisphere/Praeger.

Hammond, K. R., & Summers, D. A. (1972). Cognitive control. *Psychological Review, 79*, 58–67.

Hancock, P. A. (1986). Sustained attention under thermal stress. *Psychological Bulletin, 99*, 263–281.

Hancock, P. A. (Ed.). (1987). *Human factors psychology*. Amsterdam: Elsevier.

Hancock, P. A., & Chignell, M. H. (1985). The principle of maximal adaptability in setting stress tolerance standards. In R. E. Eberts & C. G. Eberts (Eds.), *Trends in ergonomics/human factors II*. Amsterdam: Elsevier.

Hancock, P. A., & Chignell, M. H. (1987). Adaptive control in human-machine systems. In P. A. Hancock (Ed.), *Human factors psychology* (pp. 305–345). Amsterdam: Elsevier.

Hancock, P. A., & Rosenberg, S. M. (1987). A model for evaluating stress effects of work with display units. In B. Knave & P.-G. Widebäck (Eds.), *Work with display units*. Amsterdam: Elsevier.

Hancock, P. A., & Warm, J. S. (1989). A dynamic model of stress and sustained attention. *Human Factors*, *31*, 519–537.

Harvey, L. O., Jr., Hammond, K. R., Lusk, C. M., & Mross, E. F. (1992). The application of signal detection theory to weather forecasting behavior. *Monthly Weather Review*, *120*, 863–883.

Hockey, G. R. J. (1970). Effect of loud noise on attentional selectivity. *Quarterly Journal of Experimental Psychology*, *22*, 28–36.

Hockey, G. R. J. (1986a). Changes in operator efficiency as a function of environmental stress, fatigue, and circadian rhythms. In K. R. Boff, L. Kaufman, & J. P. Thomas (Eds.), *Handbook of perception and performance: Vol. 2. Cognitive processes and performance* (pp. 44.1–44.49). New York: Wiley.

Hockey, G. R. J. (1986b). A state control theory of adaptation to stress and individual differences in stress management. In G. R. J. Hockey, A. W. K. Gaillard, & M. G. H. Coles (Eds.), *Energetics and human information processing* (pp. 285–298). Dordrecht: Nijhoff.

Hockey, R. (1979). Stress and the cognitive components of skilled performance. In V. Hamilton & D. M. Warburton (Eds.), *Human stress and cognition: An information processing approach* (pp. 141–177). Chichester: Wiley.

Hockey, R. (1983a). Current issues and new directions. In R. Hockey (Ed.), *Stress and fatigue in human performance* (pp. 363–373). Chichester: Wiley.

Hockey, R. (Ed.). (1983b). *Stress and fatigue in human performance*. Chichester: Wiley.

Hockey, R., & Hamilton, P. (1983). The cognitive patterning of stress states. In R. Hockey (Ed.), *Stress and fatigue in human performance* (pp. 331–362). Chichester: Wiley.

Hogarth, R. M. (1987). *Judgement and choice: The psychology of decision* (2nd ed.). Chichester: Wiley.

Hogarth, R. M., McKenzie, C. R. M., Gibbs, B. J., & Marquis, M. A. (1991). Learning from feedback: Exactingness and incentives. *Journal of Experimental Psychology: Learning, Memory, and Cognition*, *17*, 734–752.

Holmes, O. W. (1923). *The common law*. Boston: Little, Brown. (Original work published 1881)

Hopkins, V. D. (1988). Air traffic control. In E. L. Wiener & D. C. Nagel (Eds.), *Human factors in aviation* (pp. 639–663). San Diego: Academic Press.

Horowitz, T. (1998). Philosophical intuitions and psychological theory. *Ethics*, *108*, 367–385.

Howard, P. K. (1994). *The death of common sense: How law is suffocating America*. New York: Random House.

Hursch, C. J., Hammond, K. R., & Hursch, J. L. (1964). Some methodological considerations in multiple-cue probability studies. *Psychological Review*, *71*, 42–60.

Idzikowski, C., & Baddeley, A. D. (1983). Fear and dangerous environments. In R. Hockey (Ed.), *Stress and fatigue in human performance* (pp. 123–144). Chichester: Wiley.

Janis, I. L. (1958). *Psychological stress: Psychoanalytic and behavioral studies of surgical patients*. New York: Wiley.

Janis, I. L. (1983). Stress inoculation in health care: Theory and research. In D. Meichenbaum & M. E. Jaremko (Eds.), *Stress reduction and prevention* (pp. 67–99). New York: Plenum Press.

Janis, I., Defares, P., & Grossman, P. (1983). Hypervigilant reactions to threat. In H. Selye (Ed.), *Selye's guide to stress research* (Vol. 3, pp. 1–42). New York: Van Nostrand Reinhold.

Janis, I. L., & Mann, L. (1977). *Decision making: A psychological analysis of conflict, choice, and commitment.* New York: Free Press.

Jefferson, T. (1984). *Writings.* New York: Literary Classics of the U.S.

Jeffrey, R. (1981). *Formal logic: Its scope and limits* (2nd ed.). New York: McGraw-Hill.

Jerison, H. J. (1963). On the decrement function in human vigilance. In D. N. Buckner & J. J. McGrath (Eds.), *Vigilance: A symposium* (pp. 199–216). New York: McGraw-Hill.

Jerison, H. J. (1977). Vigilance: Biology, psychology, theory, and practice. In R. R. Mackie (Ed.), *Vigilance: Theory, operational performance, and physiological correlates* (pp. 27–41). New York: Plenum.

Jones, D. M. (1983). Noise. In R. Hockey (Ed.), *Stress and fatigue in human performance* (pp. 61–95). Chichester: Wiley.

Jorna, P. G. A. M., & Visser, R. T. B. (1991). Selection by flight simulation: Effects of anxiety on performance. In E. Farmer (Ed.), *Human resource management in aviation: Proceedings of the XVIII WEAAP conference* (Vol. 1). Aldershot, Eng.: Avebury Technical.

Kahneman, D. (1991). Judgment and decision making: A personal view. *Psychological Science, 2,* 142–145.

Kahneman, D., Slovic, P., & Tversky, A. (Eds.). (1982). *Judgment under uncertainty: Heuristics and biases.* Cambridge: Cambridge University Press.

Kahneman, D., & Tversky, A. (1972). Subjective probability: A judgment of representativeness. *Cognitive Psychology, 3,* 430–454.

Kahneman, D., & Tversky, A. (1973). On the psychology of prediction. *Psychological Review, 80,* 237–251.

Kahneman, D., & Tversky, A. (1979). Prospect theory: An analysis of decision under risk. *Econometrica, 47,* 263–291.

Kahneman, D., & Tversky, A. (1982). On the study of statistical intuitions. In D. Kahneman, P. Slovic, & A. Tversky (Eds.), *Judgment under uncertainty: Heuristics and biases* (pp. 493–508). Cambridge: Cambridge University Press.

Kaplan, M. F., Wanshula, L. T., & Zanna, M. P. (1993). Time pressure and information integration in social judgment: The effect of need for structure. In O. Svenson & A. J. Maule (Eds.), *Time pressure and stress in human judgment and decision making* (pp. 255–267). New York: Plenum.

Keeney, R. L., & Raiffa, H. (1976). *Decisions with multiple objectives: Preferences and value tradeoffs.* Cambridge: Cambridge University Press.

Keinan, G. (1987). Decision making under stress: Scanning of alternatives under controllable and uncontrollable threats. *Journal of Personality and Social Psychology, 52,* 639–644.

Keinan, G., & Friedland, N. (1984). Dilemmas concerning the training of individuals for task performance under stress. *Journal of Human Stress, 10*(4), 185–190.

Keinan, G., Friedland, N., & Ben-Porath, Y. (1987). Decision making under stress: Scanning of alternatives under physical threat. *Acta Psychologica, 64*, 219–228.

Kennan, G. F. (1993, September 30). Somalia, through a glass darkly. *New York Times*, p. A25.

Kerstholt, J. H. (1994). The effect of time pressure on decision-making behavior in a dynamic task environment. *Acta Psychologica, 84*, 89–104.

Kerstholt, J. H. (1995). Decision making in a dynamic situation: The effect of false alarms and time pressure. *Journal of Behavioral Decision Making, 8*, 181–200.

Kirschenbaum, S. S. (1992). Influence of experience on information-gathering strategies. *Journal of Applied Psychology, 77*, 343–352.

Kirschenbaum, S. S., & Arruda, J. E. (1994). Effects of graphic and verbal probability information on command decision making. *Human Factors, 36*, 406–418.

Klapp, S. T. (1987). Short-term memory limits in human performance. In P. A. Hancock (Ed.), *Human factors psychology* (pp. 1–27). Amsterdam: Elsevier.

Kleinmuntz, D. N. (1987). Human decision processes: Heuristics and task structure. In P. A. Hancock (Ed.), *Human factors psychology* (pp. 123–157). Amsterdam: Elsevier.

Koch, S. (Ed.). (1959–1963). *Psychology: A study of a science* (Vol. 1–6). New York: McGraw-Hill.

Koehler, J. J. (1996). The base rate fallacy reconsidered: Descriptive, normative, and methodological challenges. *Behavioral and Brain Sciences, 19*, 1–17, 41–53.

Koelega, H. S., & Brinkman, J.-A. (1986). Noise and vigilance: An evaluative review. *Human Factors, 28*, 465–481.

Koelega, H. S., Brinkman, J.-A., & Bergman, H. (1986). No effect of noise on vigilance performance? *Human Factors, 28*, 581–593.

Krueger, G. P., Armstrong, R. N., & Cisco, R. R. (1985). Aviator performance in week-long extended flight operations in a helicopter simulator. *Behavior Research Methods, Instruments, & Computers, 17*, 68–74.

Lazarus, R. S. (1966). *Psychological stress and the coping process.* New York: McGraw-Hill.

Lazarus, R. S. (1990). Theory-based stress measurement. *Psychological Inquiry, 1*(1), 3–13.

Lazarus, R. S. (1991). *Emotion and adaptation.* New York: Oxford University Press.

Lazarus, R. S., & Cohen, J. (1977). Environmental stress. In J. Wohlwill & I. Altman (Eds.), *Human behavior and environment* (pp. 90–127). New York: Plenum.

Lazarus, R. S., & Folkman, S. (1984). *Stress, appraisal, and coping.* New York: Springer.

Lazarus, R. S., & Folkman, S. (1987). Transactional theory and research on emotions and coping. *European Journal of Personality, 1*, 141–169.

Lazarus, R. S., & Lazarus, B. N. (1994). *Passion and reason: Making sense of our emotions.* New York: Oxford University Press.

LeDoux, J. E. (1996). *The emotional brain: The mysterious underpinnings of emotional life.* New York: Simon & Schuster.

Lee, A. T., & Bussolari, S. R. (1989). Flight simulator platform motion and air transport pilot training. *Aviation, Space, and Environmental Medicine, 60,* 136–140.

Levi, A., & Tetlock, P. E. (1980). A cognitive analysis of Japan's 1941 decision for war. *Journal of Conflict Resolution, 24,* 195–211.

Levine, S. (1990). *Stress and performance.* Washington, DC: National Academy Press.

Lichtenstein, S., Slovic, P., Fischhoff, B., Layman, M., & Combs, B. (1978). Judged frequency of lethal events. *Journal of Experimental Psychology: Human Learning and Memory, 4,* 551–578.

Liu, Y., & Wickens, C. D. (1994). Mental workload and cognitive task automaticity: An evaluation of subjective and time estimation metrics. *Ergonomics, 37,* 1834–1854.

Llewellyn-Thomas, H., & O'Connor, A. (1998). Improving evidence-based decision making: Recent advances in research-transfer models. *Medical Decision Making, 18,* 1–18.

Lopes, L. L. (1976). Model-based decision and inference in stud poker. *Journal of Experimental Psychology: General, 105,* 217–239.

Lopes, L. L. (1991). The rhetoric of irrationality. *Theory & Psychology, 1*(1), 65–82.

Lopes, L. L., & Oden, G. C. (1991). The rationality of intelligence. In E. Eells & T. Maruszewski (Eds.), *Probability and rationality: Studies on L. Jonathan Cohen's philosophy of science* (pp. 199–223). Amsterdam: Rodopi.

Luce, M. F., Bettman, J. R., & Payne, J. W. (1997). Choice processing in emotionally difficult decisions. *Journal of Experimental Psychology: Learning, Memory, and Cognition, 23,* 384–405.

Lusk, C. M. (1993). Assessing components of judgments in an operational setting: The effects of time pressure on aviation weather forecasting. In O. Svenson & A. J. Maule (Eds.), *Time pressure and stress in human judgment and decision making* (pp. 309–321). New York: Plenum.

Maclean, N. (1992). *Young men and fire.* Chicago: University of Chicago Press.

Mahan, R. P. (1991). The effects of extended work on multi-dimensional judgments of a manufacturing process. *International Journal of Human Factors in Manufacturing, 1,* 155–165.

Mahan, R. P. (1992). Effects of task uncertainty and continuous performance on knowledge execution in complex decision-making. *International Journal of Computer Integrated Manufacturing, 5*(2), 58–67.

Mahan, R. P. (1994). Stress-induced strategy shifts toward intuitive cognition: A cognitive continuum framework approach. *Human Performance, 7,* 85–118.

Malaterre, G., Ferrandez, F., Fleury, D., & Lechner, D. (1988). Decision making in emergency situations. *Ergonomics, 31,* 643–655.

Mandler, G. (1982). Stress and thought processes. In L. Goldberger & S. Breznitz (Eds.), *Handbook of stress: Theoretical and clinical aspects* (pp. 88–104). New York: Free Press.

Mann, L., & Janis, I. (1982). Conflict theory of decision making and the expectancy-value approach. In N. T. Feather (Ed.), *Expectations and actions: Expectancy-value models in psychology* (pp. 341–364). Hillsdale, NJ: Erlbaum.

Margolis, H. (1987). *Patterns, thinking, and cognition: A theory of judgment.* Chicago, IL: University of Chicago Press.

Massaro, D. W., & Friedman, D. (1990). Models of integration given multiple sources of information. *Psychological Review, 97*, 225–252.

May, E. R., & Zelikow, P. D. (Eds.). (1997). *The Kennedy tapes: Inside the White House during the Cuban missile crisis.* Cambridge: Belknap Press of Harvard University Press.

McEwen, B. S. (1998). Seminars in medicine of the Beth Israel Deaconess Medical Center: Protective and damaging effects of stress mediators. *New England Journal of Medicine, 338*, 171–179.

McEwen, B. S., & Stellar, E. (1993). Stress and the individual: Mechanisms leading to disease. *Archives of Internal Medicine, 153*, 2093–2101.

McGrath, J. J. (1963). Some problems of definition and criteria in the study of vigilance performance. In D. N. Buckner & J. J. McGrath (Eds.), *Vigilance: A symposium* (pp. 227–246). New York: McGraw-Hill.

McPherson, J. (1997). *For cause and comrades: Why men fought in the Civil War.* New York: Oxford University Press.

Mellers, B. A., Schwartz, A., Ho, K., & Ritov, L. (1997). Decision affect theory: Emotional reactions to the outcomes of risky options. *Psychological Science, 8*, 423–429.

Melton, C. E. (1982). *Physiological stress in air traffic controllers: A review* (FAA-AM-82–17). Washington, DC: U.S. Department of Transportation, Federal Aviation Administration.

Milgram, S. (1974). *Obedience to authority.* New York: Harper & Row.

Miller, G. A. (1956). The magical number seven, plus or minus two: Some limits on our capacity for processing information. *Psychological Review, 63*, 81–97.

Monroe, S., & Kelley, J. (1995). Measurement of stress appraisal. In S. Cohen, R. C. Kessler, & L. U. Gordon (Eds.), *Measuring stress: A guide for health and social scientists* (pp. 122–147). New York: Oxford University Press.

Nasby, W. (1996). Moderators of mood-congruent encoding and judgement: Evidence that elated and depressed moods implicate distinct processes. *Cognition and Emotion, 10*, 361–377.

Newell, A., & Simon, H. A. (1972). *Human problem solving.* Englewood Cliffs, NJ: Prentice-Hall.

O'Brien, C. (1996). *The long affair: Thomas Jefferson and the French revolution 1785–1800.* Chicago: University of Chicago Press.

OSS Assessment Staff. (1948). *Assessment of men: Selection of personnel for the Office of Strategic Services.* New York: Rinehart.

Parkes, K. R. (1984). Locus of control, cognitive appraisal, and coping in stressful episodes. *Journal of Personality and Social Psychology, 46*, 655–668.

Paterson, R. J., & Neufeld, R. W. J. (1987). Clear danger: Situational determinants of the appraisal of threat. *Psychological Bulletin, 101*, 404–416.

Paulen, M. (1887). La simultanéité des actes psychiques. *Revue Scientifique, 39*, 684–689.

Payne, J. W. (1976). Task complexity and contingent processing in decision making: An information search and protocol analysis. *Organizational Behavior and Human Performance, 16*, 366–387.

Payne, J. W. (1980). Information processing theory: Some concepts and methods applied to decision research. In T. S. Wallsten (Ed.), *Cognitive processes in choice and decision behavior* (pp. 95–115). Hillsdale, NJ: Erlbaum.

Payne, J. W. (1982). Contingent decision behavior. *Psychological Bulletin, 92*, 382–402.

Payne, J. W., Bettman, J. R., & Johnson, E. J. (1988). Adaptive strategy selection in decision making. *Journal of Experimental Psychology: Learning, Memory, and Cognition, 14*, 534–552.

Payne, J. W., Bettman, J. R., & Luce, M. F. (1996). When time is money: Decision behavior under opportunity-cost time pressure. *Organizational Behavior and Human Decision Processes, 66*, 131–152.

Payne, J. W., & Braunstein, M. L. (1978). Risky choice: An examination of information acquisition behavior. *Memory & Cognition, 6*, 554–561.

Payne, J. W., Johnson, E. J., Bettman, J. R., & Coupey, E. (1989). *Understanding contingent choice: A computer simulation approach.* Unpublished manuscript.

Pennington, N., & Hastie, R. (1991). A cognitive theory of juror decision making: The story model. *Cardozo Law Review, 13*, 519–557.

Pennington, N., & Hastie, R. (1993). The story model for juror decision making. In R. Hastie (Ed.), *Inside the juror: The psychology of juror decision making* (pp. 192–221). Cambridge: Cambridge University Press.

Plous, S. (1993). *The psychology of judgment and decision making.* Philadelphia: Temple University Press.

Poulton, E. C. (1976a). Arousing environmental stresses can improve performance, whatever people say. *Aviation, Space, and Environmental Medicine, 47*, 1193–1204.

Poulton, E. C. (1976b). Continuous noise interferes with work by masking auditory feedback and inner speech. *Applied Ergonomics, 7*, 79–84.

Poulton, E. C. (1994). *Behavioral decision theory: A new approach.* New York: Cambridge University Press.

Presidential Commission on the Space Shuttle Challenger Accident. (1986). *Report of the Presidential Commission on the Space Shuttle Challenger Accident.* Washington, DC: Author.

Raby, M., & Wickens, C. D. (1994). Strategic workload management and decision biases in aviation. *The International Journal of Aviation Psychology, 4*, 211–240.

Raby, M., Wickens, C. D., & Marsh, R. (1990). *Investigation of factors comprising a model of pilot decision making: Part I. Cognitive biases in workload management strategy* (ARL-90-7/SCEEE-90-1). Urbana-Champaign: University of Illinois, Aviation Research Laboratory.

Raiffa, H. (1968). *Decision analysis: Introductory lectures on choices under uncertainty.* Reading, MA: Addison-Wesley.

Raiffa, H. (1969). *Preferences for multiattributed alternatives* (Report No. RM-5868–DOT/RC). Santa Monica, CA: Rand Corporation.

Ramsey, J. D., Burford, C. L., Beshir, M. Y., & Jensen, R. C. (1983). Effects of workplace thermal conditions on safe work behavior. *Journal of Safety Research, 14*, 105–114.

Rasmussen, J., & Rouse, W. B. (Eds.). (1981). *Human detection and diagnosis of system failures.* New York: Plenum Press.

Raven, J. C. (1962). *Advanced progressive matrices, set II.* London: H. K. Lewis.

Rothstein, H. G. (1986). The effects of time pressure on judgment in multiple cue probability learning. *Organizational Behavior and Human Decision Processes, 37*, 83–92.

Saegert, S., & Winkel, G. H. (1990). Environmental psychology. *Annual Review of Psychology, 41*, 441–477.

Salamé, P., & Wittersheim, G. (1978). Selective noise disturbance of the information input in short-term memory. *Quarterly Journal of Experimental Psychology, 30*, 693–704.

Sanders, A. F. (1961). The influence of noise on two discrimination tasks. *Ergonomics, 4*, 253–259.

Savage, L. J. (1954). *The foundations of statistics.* New York: Wiley.

Schwartz, D. R., & Howell, W. C. (1985). Optional stopping performance under graphic and numeric CRT formatting. *Human Factors, 27*, 433–444.

Serfaty, D., Entin, E. E., & Volpe, C. (1993). Adaptation to stress in team decision-making and coordination. In *Proceedings of Human Factors and Ergonomics Society 37th Annual Meeting* (Vol. 2, pp. 1228–1232). Santa Monica, CA: Human Factors and Ergonomics Society.

Shanteau, J. (1974). Component processes in risky decision making. *Journal of Experimental Psychology, 103*, 680–691.

Shanteau, J., & Dino, G. A. (1983). *Stress effects on problem-solving and decision-making behavior.* Paper presented at the twenty-fourth annual meeting of the Psychonomic Society, San Diego. (From *Bulletin of the Psychonomic Society, 21,* 362, Abstract No. 309).

Shanteau, J., & Dino, G. A. (1993). Environmental stressor effects on creativity and decision making. In O. Svenson & A. J. Maule (Eds.), *Time pressure and stress in human judgment and decision making* (pp. 293–308). New York: Plenum.

Sheridan, T. B. (1981). Understanding human error and aiding human diagnostic behavior in nuclear power plants. In J. Rasmussen & W. Rouse (Eds.), *Human detection and diagnosis of system failures* (pp. 19–36). New York: Plenum.

Simon, H. A. (1955). A behavioral model of rational choice. *Quarterly Journal of Economics, 69*, 99–118.

Simon, H. A. (1956). Rational choice and the structure of the environment. *Psychological Review, 63*, 129–138.

Simon, H. A. (1979). *Models of thought.* New Haven, CT: Yale University Press.

Slovic, P., & Lichtenstein, S. (1971). Comparison of Bayesian and regression approaches to the study of information processing in judgment. *Organizational Behavior and Human Performance, 6*, 649–744.

Smith, A. P., Jones, D. M., & Broadbent, D. E. (1981). The effects of noise on recall of categorized lists. *British Journal of Psychology, 72*, 299–316.

Smith, J. F., & Kida, T. (1991). Heuristics and biases: Expertise and task realism in auditing. *Psychological Bulletin, 109*, 472–489.

Smith, R. C. (1985). Stress, anxiety, and the air traffic control specialist: Some surprising conclusions from a decade of research. In C. D. Spielberger, I. G. Sarason, & P. B. Defares (Eds.), *Stress and anxiety* (Vol. 9, pp. 95–109). Washington, DC: Hemisphere.

Smith, S. (1997). *The Constitution and the pride of reason.* New York: Oxford University Press.

Sperling, P., & Eyer, J. (1988). "Allostasis": A new paradigm to explain arousal pathology. In R. Fisher & J. Reason (Eds.), *Handbook of life stress, cognition, and health* (pp. 629–652). New York: Wiley.

Stewart, T. R., & Lusk, C. M. (1994). Seven components of judgmental forecasting skill: Implications for research and the improvement of forecasts. *Journal of Forecasting, 13*, 579–599.

Stiensmeier-Pelster, J., & Schürmann, M. (1993). Information processing in decision making under time pressure: The influence of action versus state orientation. In O. Svenson & A. J. Maule (Eds.), *Time pressure and stress in human judgment and decision making* (pp. 241–253). New York: Plenum.

Stokes, A. F. (1991). MIDIS—A microcomputer flight decision simulator. In E. Farmer (Ed.), *Human resource management in aviation: Proceedings of the XVIII WEAAP conference* (Vol. 1, pp. 107–121). Aldershot, Eng.: Avebury Technical.

Stokes, A., Belger, A., & Zhang, K. (1990). *Investigation of factors comprising a model of pilot decision making: Part II. Anxiety and cognitive strategies in expert and novice aviators* (Tech. Rep. No. ARL-90-8/SCEEE-90-2). Savoy: University of Illinois, Aviation Research Laboratory, Institute of Aviation.

Svenson, O., & Benson, L., III (1993a). Framing and time pressure in decision making. In O. Svenson & A. J. Maule (Eds.), *Time pressure and stress in human judgment and decision making* (pp. 133–144). New York: Plenum.

Svenson, O., & Benson, L., III (1993b). On experimental instructions and the inducement of time pressure behavior. In O. Svenson & A. J. Maule (Eds.), *Time pressure and stress in human judgment and decision making* (pp. 157–165). New York: Plenum.

Svenson, O., & Maule, A. J. (Eds.). (1993). *Time pressure and stress in human judgment and decision making.* New York: Plenum.

Swets, J. A. (1973). The relative operating characteristic in psychology. *Science, 182*, 990–1000.

Taylor, S. E. (1991). Asymmetrical effects of positive and negative events: The mobilization-minimization hypothesis. *Psychological Bulletin, 110*, 67–85.

Thomas, J. (1997, June 4). Bomb trial judge warns both sides against "lynching." *New York Times,* pp. A1, 6.

Thompson, M. M., Naccarato, M. E., & Parker, K. H. (1989, June). *Measuring cognitive needs: The development of the personal need for structure (PNS) and personal fear of invalidity (PFI) scales.* Paper presented at the annual meeting of the Canadian Psychological Association, Halifax.

Thurstone, L. L. (1931). The measurement of change in social attitudes. *Journal of Social Psychology, 2*, 230–235.

Tooby, J., & Cosmides, L. (1992). The psychological foundations of culture. In J. H. Barkow, L. Cosmides, & J. Tooby (Eds.), *The Adapted mind: Evolutionary psychology and the generation of culture* (pp. 19–136). New York: Oxford University Press.

Tucker, L. R. (1964). A suggested alternative formulation in the developments by Hursch, Hammond, and Hursch, and by Hammond, Hursch, and Todd. *Psychological Review, 71*, 528–530.

Tversky, A. (1972). Elimination by aspects: A theory of choice. *Psychological Review, 79*, 281–299.

Tversky, A., & Kahneman, D. (1974). Judgment under uncertainty: Heuristics and biases. *Science, 185*, 1124–1131.

Tversky, A., & Kahneman, D. (1981). The framing of decisions and the psychology of choice. *Science, 211*, 453–458.

Tversky, A., & Kahneman, D. (1983). Extensional versus intuitive reasoning: The conjunction fallacy in probability judgment. *Psychological Review, 90*, 293–315.

Tversky, A., Sattath, S., & Slovic, P. (1988). Contingent weighting in judgment and choice. *Psychological Review*, *95*, 371–384.

Verplanken, B. (1993). Need for cognition and external information search: Responses to time pressure during decision-making. *Journal of Research in Personality*, *27*, 238–252.

Vessey, I. (1991). Cognitive fit: A theory-based analysis of the graphs versus tables literature. *Decision Sciences*, *22*, 219–240.

Vicente, K. (1992). Multilevel interfaces for power plant control rooms. II: A preliminary design space. *Nuclear Safety*, *33*, 543–548.

von Neumann, J., & Morgenstern, O. (1947). *Theory of games and economic behavior* (2nd ed.). Princeton, NJ: Princeton University Press.

von Winterfeldt, D., & Edwards, W. (1986). *Decision analysis and behavioral research.* Cambridge: Cambridge University Press.

Wallsten, T. S. (1993). Time pressure and payoff effects on multidimensional probabilistic inference. In O. Svenson & A. J. Maule (Eds.), *Time pressure and stress in human judgment and decision making* (pp. 167–179). New York: Plenum.

Wells, A., & Matthews, G. (1994). Self-consciousness and cognitive failures as predictors of coping in stressful episodes. *Cognition and Emotion*, *8*, 279–295.

Weltman, G., Smith, J. E., & Egstrom, G. H. (1971). Perceptual narrowing during simulated pressure-chamber exposure. *Human Factors*, *13*, 99–107.

Wertheimer, M. (1998). Opus Magnificentissimum. *Contemporary Psychology*, *43*, 7–10.

Wescott, L. & Degen, P. (1983). *Wind and Sand: The Story of the Wright Brothers at Kitty Hawk.* Eastern Acorn Press, Eastern National Park and Monument Association.

Wickens, C. D. (1987). Attention. In P. A. Hancock (Ed.), *Human factors psychology* (pp. 29–80). Amsterdam: Elsevier.

Wickens, C. D. (1996). Designing for stress. In J. E. Driskell & E. Salas (Eds.), *Stress and human performance* (pp. 279–295). Mahwah, NJ: Erlbaum.

Wickens, C. D., & Flach, J. (1988). Human information processing. In E. L. Wiener & D. C. Nagel (Eds.), *Human factors in aviation* (pp. 111–155). San Diego: Academic Press.

Wickens, C. D., Stokes, A., Barnett, B., & Hyman, F. (1988). *The effects of stress on pilot judgement in a MIDIS simulator* (Tech. Rep. No. ARL-88-5/INEL-88-1). Urbana-Champaign: University of Illinois, Aviation Research Laboratory, Institute of Aviation.

Wickens, C. D., Stokes, A., Barnett, B., & Hyman, F. (1993). The effects of stress on pilot judgment in a MIDIS simulator. In O. Svenson & A. J. Maule (Eds.), *Time pressure and stress in human judgment and decision making* (pp. 271–292). New York: Plenum.

Wiener, E. L., & Nagel, D. C. (Eds.). (1988). *Human factors in aviation.* San Diego: Academic Press.

Wills, G. (1978). *Inventing America: Jefferson's Declaration of Independence.* Garden City, NY: Doubleday.

Wilson, J. Q. (1993). *The moral sense.* New York: Free Press.

Wokoun, W. (1969). Music for working. *Science Journal*, *5A*(5), 54–59.

Woodhead, M. M. (1964). Searching a visual display in intermittent noise. *Journal of Sound and Vibration*, *1*, 157–161.

Wright, P. (1974). The harassed decision maker: Time pressures, distractions, and the use of evidence. *Journal of Applied Psychology, 59*, 555–561.
Xiao, Y., Hunter, W. A., Mackenzie, C. F., Jefferies, N. J., & Horst, R. L. (1996). Task complexity in emergency medical care and its implications for team coordination. *Human Factors, 38*, 636–645.
Yates, J. F. (1990). *Judgment and decision making*. Englewood Cliffs, NJ: Prentice-Hall.
Zajonc, R. B. (1980). Feeling and thinking: Preferences need no inferences. *American Psychologist, 35*, 151–175.
Zakay, D. (1985). Post-decisional confidence and conflict experienced in a choice process. *Acta Psychologica, 58*, 75–80.
Zakay, D., & Wooler, S. (1984). Time pressure, training and decision effectiveness. *Ergonomics, 27*, 273–284.
Zsambok, C. E., & Klein, G. A. (1997). *Naturalistic decision making*. Mahwah, NJ: Erlbaum.

Index